Consumer trends and new product opportunities in the food sector

Consumer trends and new product opportunities in the food sector

edited by: Klaus G. Grunert

Wageningen Academic
P u b l i s h e r s

EAN: 9789086863075
e-EAN: 9789086868520
ISBN: 978-90-8686-307-5
e-ISBN: 978-90-8686-852-0
DOI: 10.3920/978-90-8686-852-0

Cover design: Izabela Jasiczak

First published, 2017

© Wageningen Academic Publishers
The Netherlands, 2017

Wageningen Academic Publishers,
P.O. Box 220, 6700 AE Wageningen,
the Netherlands,
www.WageningenAcademic.com
copyright@WageningenAcademic.com

This book has been co-funded by the Erasmus+ programme of the European Union 'Food Quality & Consumer Studies' (Strategic partnership Erasmus + Nr. 2014-1-SK01-KA203-000464).

The European Commission support for the production of this publication does not constitute an endorsement of the contents which reflects the views only of the authors, and the Commission cannot be held responsible for any use which may be made of the information contained therein.

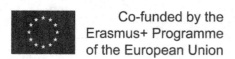

Book reviewers:
Miroslava Kačániová
Slovak University of Agriculture in Nitra, Slovak Republic
Simona Kunová
Slovak University of Agriculture in Nitra, Slovak Republic

Table of contents

Preface

This book is based on qualified contributions of experts in the field of food marketing and consumer studies. All the chapters are outcomes of international project cooperation within Erasmus+ Strategic partnership project 'Food Quality & Consumer Studies' started in 2014 with the aim to modernize and improve the quality of university education in the field of food science, food marketing and consumer studies, applied thorough the synergic effect of international cooperation, transfer of innovation and creation of new values in project consortium. Results of the projects account 4 intellectual outputs in forms of the books and concept of the summer schools and training in specific fields related to food industry and market.

Consumer needs and wants should drive marketing decisions in the company so to know and understand ongoing and expected consumer trends on food market belong to inevitable part of decision making processes. I believe that this book "Consumer Trends and New Product Opportunities in the Food Sector" will provide readers from academic community and business sphere both with theoretical knowledge and latest viewpoints necessary for understanding, analysing and implementing the latest food trends in product research and development, production, marketing and communication.

Publishing of the book has been co-funded by the Erasmus+ programme of the European Union 'Food Quality & Consumer Studies' (Nr. 2014-1-SK01-KA203-000464). I would like to thank all project team members, especially book editor, prof. Klaus Grunert and authors of different chapters, and also to reviewers as well as all those who in any way contributed to the content or formal aspect of the book.

Elena Horská
Project coordinator

1. Introduction

K.G. Grunert

Aarhus University, Fuglesangsalle 4, 8210 Aarhus V, Denmark; klg@mgmt.au.dk

Like in all other industries, food producers need to innovate and develop new products to bring to the market. As existing products reach the peak of their lifecycle and competitors catch up on any competitive advantage there may have been, innovation by new product development is the way to ensure future growth. However, it is well known that innovation in the food industry is risky, with the majority of new product introductions failing on the market. In other words, innovation in the food industry is both necessary and risky.

There are two major sources for new product development: technological advances and changing customer needs. While the food industry has been traditionally viewed as low-tech compared to other industries, and while much of the technological advances that have been made over the years have been aimed at process optimization and not so much at developing new products, this situation is changing. The focus on biochemistry and biotechnology in recent years has opened up for new ways of producing food, both in primary production and in food processing. These are foods with improved sensory properties, with improved health characteristics and with new forms of convenience. These are also foods that are better tailored to the different needs of consumer segments. These are foods that are produced in a way that consumers find transparent, sustainable and authentic.

Because consumer needs and wants with regard to food are changing. Some of these changes have been in the making for many years, others are of more recent origin. For several decades we have seen a health trend building up, where healthfulness becomes an important motive for consumers that competes with the basic motive of seeking pleasure and sensory reward in food. Likewise, we have seen for many years the trends towards more convenience, with time-pressured consumers looking for ways to save time and effort in the shopping, preparation, and consumption of food. While the demand for convenience earlier on was mostly linked to people not so interested in food and to food of low quality, we see now an increasing demand for convenience products of high quality also by people for whom food is important. More recently, we have seen the development of a sharpened focus on the authenticity of food. Perhaps as a reaction to the globalization of food supply and increasingly less transparent food chains, many consumers focus on products that they regard as real, natural, unalienated, unspoilt, original, artisanal and from short supply chains. And we start seeing a focus on the sustainability of food production, with at least some people starting to pay attention to topics like the environmental impact of meat production, food miles, and alternative sources of protein.

For the food industry, these developments carry considerable potential, but also considerable risk. Foods with specific health properties, the so-called functional foods, have been developed, but sometimes fail because consumers don't regard them as authentic. Convenience products fail because manufacturers misperceive consumers' sensory preferences or consumers quickly become bored of them. Artisanal products fail because even an artisanal producer needs to live up to the requirements of modern retail chains if they want to achieve wider distribution. Innovation in the food sector these days involves lots of trade-offs and an intimate understanding of the way in which consumers form preferences.

In this book, we will look at consumer trends, at trends in food production, and at where these two types of trends should come together – new product development in the food sector. In the first part of the book, we look at the four trends already briefly mentioned above – the health trend, the sustainability trend, the authenticity trend and the convenience trend. In the second part, we look at trends on the supply side. The supply side is the whole food chain, and we therefore will look both at the entire food chain and its major actors, namely farmers, food processors and retailers. We have a special chapter on the modernization of traditional foods, as an important way to accommodate the authenticity trend. In the last part, we look at new product development in food sector and deal with the questions of segmentation, the global-local issue, and success factors in new product development in the food sector.

This book is one of the outcomes of the FOODCOST project, funded by the EU under the Erasmus+ program, agreement number 2014-1-SK01-KA203-000464, coordinated by professor Elena Horská from the Slovak University of Agriculture in Nitra. We hope that it will contribute to successful innovation in the food sector.

2. The health trend

K.G. Grunert

Aarhus University, Fuglesangsalle 4, 8210 Aarhus V, Denmark; klg@mgmt.au.dk

Abstract

Food and health has been an issue of growing importance on the public agenda for many years. As a result, many consumers have learned about the relationship between food properties and health, and many manufacturers have launched products with a health positioning. However, it is necessary to understand consumer decision-making when buying food in order to profit from the health trend. Many consumer food choices are habitual and based on simplified ways of decision-making, and health will often not be the only and not the main motive for their choices. In addition, consumers differ in how they deal with the food and health issue. While some consumers respond positively to products with specific health properties, others judge a product's healthfulness in a more holistic way. Communicating the healthfulness of a food product to consumers therefore needs to be adapted to different ways in which consumers handle the food and health issue.

Keywords: healthfulness, consumer decision-making, health communication, health claims, nutrition

2.1 Introduction

Today, everybody knows that what we eat has an impact on our health. However, the role of health in both consumers' food choice and in the food industry's product development has not always been that prominent. While in some parts of the world, and notably in parts of Southeast Asia, there is a long tradition for using food and food ingredients to manage health and well-being, in most of the Western world the issue of food and health is of much more recent origin. While no exact date can be set, it was in the post-world-war II years that a focus on non-communicable diseases and later on overweight and obesity resulted in growing awareness of the link between food and health, first by public authorities and other actors dealing with public health, then by the food industry and, much helped by the media, eventually by the public.

Today, we can look back at several decades of educational and informational activities aimed at informing people about the consequences of an unhealthy diet and the elements of a healthy diet. For all we know, this has had only limited effect on people's dietary habits (Capaci *et al.*, 2012). The reasons for this will be discussed in more detail below, but briefly speaking it has to do with people's motives when choosing food and with the way in which food choices are made. Health is obviously important for everyone, but is not the only and usually not even the most important criterion when choosing food – normally pleasure, taste and family liking come first (Grunert and Wills, 2007). People may therefore perceive trade-offs between health and other motives, notably taste and pleasure. This issue is exacerbated by the way in which people usually food, which is not conducive to involved trade-offs between different buying motives. Much of food choice is habitual, intuitive, not well-reasoned and fast (Wansink and Sobal, 2007). Therefore, many times the dominant motive will prevail. So many people continue to make food choices that, from a health perspective, are not optimal, while at the same time feeling the pressure to eat more healthily that has characterized the public arena for many years (Chrysochou *et al.*, 2010).

This situation has created opportunities for the food industry. There has been a proliferation of products positioned and marketed as healthy. This includes products for the consumer segment that indeed does manage their dietary intake from a nutritional perspective, including consumers that have developed specific preferences for or against certain food ingredients, like gluten or lactose, that are not necessarily in line with mainstream nutritional wisdom. It includes products that have been enriched or modified to have specific health benefits, the so-called functional foods, where the health benefit is usually communicated in form of a health claim. It includes products marketed as free from additives or chemicals, because many consumers believe that this makes products healthier. The success of organic food products is partly due to consumers' conviction that these are healthier than conventional products (Thøgersen, 2009), even though it is much debated whether this is indeed the case. Healthy food products has become big business, but many food products launched with a health positioning have also failed in the market. As already hinted at, consumers' approach to the food and health issue is ambivalent, and this ambivalence also characterizes consumer acceptance, or lack of it, of food with a health positioning. A good understanding of how the health aspect enters consumer decision-making is therefore a prerequisite for the successful development and marketing of food products that profit from the health trend.

In the present chapter, we will first start by looking at what consumers indeed know about food and health, and will then discuss when and to which extent this knowledge turns into food choices. We will then look at ways in which the healthfulness of a product can be communicated to consumer, and finally at how consumers deal with the trade-off between health and pleasure. On this background, we will then assess the possibilities for successful new product development profiting from the health trend.

2.2 What do people know about healthy eating?

In order to make healthy food choices, consumers must have some knowledge on what constitutes healthy eating and healthful products. This knowledge can guide the composition of a diet, the composition of a meal, and the choice of products both across and within product categories. Most notably, it is a prerequisite for understanding health communication about food products, especially from manufacturers, including information that is on the food label.

This means that, in order incorporate healthfulness into their shopping scripts, shoppers need to be aware of nutrition recommendations and basic food-based guidelines. They should also be able to apply this knowledge in their food choices and eating behaviour (Sapp and Jensen, 1997). Many studies have reported a positive association between nutrition knowledge and healthful food behaviour (Dallongeville *et al.*, 2001; Handu *et al.*, 2008; Klohe-Lehman *et al.*, 2006; Lee *et al.*, 2009; Wardle *et al.*, 2000). People's nutritional knowledge can be measured by conducting knowledge tests, and a number of instruments for doing this are available (e.g. Grunert *et al.*, 2012; Parmenter and Wardle, 1999). Table 2.1 shows results from one such study, measuring respondents knowledge of dietary recommendations, sources of nutrients, and knowledge on the calorie content of food and drink products in six European countries (Grunert *et al.*, 2012).

Table 2.1. Nutrition knowledge in six European countries. Correct answers are in bold. All figures are percentages of respondents, n=5,967 in UK, Sweden, France, Germany, Poland, Hungary (Grunert *et al.*, 2012).

Health experts recommend to eat...

	More	About the same	Less	Try to avoid	Don't understand what it means
Fat	1.7	7.1	**63.4**	25.9	0.3
Polyunsaturated fats	**25.4**	19.7	24.7	16.7	10.1
Calories	2.9	20.4	**64.6**	8.9	0.6
Sodium	3.2	17.3	**44.6**	15.9	14.2
Saturated fat	3.9	11.0	**37.6**	37.9	5.9
Whole grains	**77.8**	15.4	3.3	1.1	0.4
Salt	1.1	8.1	**65.8**	22.6	0.4
Trans fat	1.8	6.8	**25.2**	36.5	25.2
Sugar	1.3	5.5	**66.7**	23.7	0.5
Omega-3 fatty acids	**61.7**	18.1	6.8	3.4	7.6
Fibre	**77.2**	16.1	3.3	1.0	0.9
Monounsaturated fat	11.1	**25.8**	27.5	9.9	22.1

Health experts recommend to eat...

	A lot	Some	A little	Try to avoid	Not answered
Fruits and vegetables	**96.5**	2.8	0.3	0.1	0.4
Starchy foods (bread, rice, pasta, potatoes)	**14.6**	62.1	20.7	1.5	1.0
Protein sources	39.4	**55.3**	4.1	0.2	1.0
Milk and dairy products	35.1	**53.9**	9.1	0.5	1.3
Foods and drinks that are high in fat	1.1	5.9	**41.2**	50.5	1.3
Foods & drinks high in sugars	1.0	3.7	**33.3**	60.7	1.3
Foods and drinks that are high in salt	0.7	4.7	**33.9**	59.6	1.1

Health experts recommend to eat...

	High	Low	Not answered
Fat – walnuts	**68.8**	26.8	4.5
Fat – bread	11.9	**79.0**	9.0
Fat – margarine	**75.8**	18.5	5.8

Table 2.1. Continued.

	High	Low	Not answered
Fat – corn flakes	14.4	**75.5**	10.1
Fat – chocolate	**79.9**	14.7	5.4
Fat – avocado	**47.0**	45.4	7.7
Fat – carp/cod[1]	29.2	**62.9**	7.9
Fat – olive oil	**64.0**	28.4	7.6
Fat – regular yoghurt	26.4	**67.1**	6.5
Fat – smoked salmon[2]	**57.0**	35.8	7.1
Fat – red meat	**42.8**	50.5	6.8
Saturated fat – walnuts	35.3	**50.0**	14.8
Saturated fat – bread	13.5	**70.4**	16.0
Saturated fat – margarine	59.5	**28.8**	11.7
Saturated fat – corn flakes	14.5	**69.4**	16.1
Saturated fat – chocolate	**56.0**	28.9	15.0
Saturated fat – avocado	23.6	**61.3**	15.1
Saturated fat – cod[1]	26.2	**59.8**	14.1
Saturated fat – olive oil	38.3	**49.1**	12.6
Saturated fat – regular yoghurt	**18.7**	65.9	15.4
Saturated fat – smoked salmon[2]	35.8	**50.3**	13.9
Saturated fat – red meat	**38.1**	46.6	15.3
Salt – walnuts	6.6	**78.7**	14.6
Salt – bread	**31.8**	60.5	7.7
Salt – margarine	**30.2**	57.6	12.2
Salt – corn flakes	**15.3**	71.6	13.1
Salt – chocolate	6.0	**78.1**	15.9
Salt – avocado	4.8	**78.1**	17.2
Salt – cod[1]	17.3	**67.1**	15.7
Salt – olive oil	8.0	**74.5**	17.4
Salt – regular yoghurt	5.0	**78.4**	16.5
Salt – smoked salmon[2]	47.2	40.4	12.4
Salt – red meat	20.2	**63.7**	16.1
Sugar – walnuts	5.4	**79.6**	15.0
Sugar – bread	19.9	**68.8**	11.3
Sugar – margarine	5.6	**77.2**	17.2

Table 2.1. Continued.

	High	Low	Not answered
Sugar – corn flakes	48.5	**43.4**	8.1
Sugar – chocolate	**87.7**	6.4	5.9
Sugar – avocado	18.7	**65.7**	15.6
Sugar – cod[1]	2.3	**79.3**	18.3
Sugar – olive oil	3.3	**78.1**	18.6
Sugar – regular yoghurt	29.3	**59.4**	11.3
Sugar – regular yoghurt	3.9	**77.8**	18.3
Sugar – red meat	3.6	**77.4**	19.0

How many calories are in the following product?

	Up to 40 calories	41-100 calories	101-200 calories	201-300 calories	301-400 calories	401-480 calories	481 calories or more	Not answered
330 ml can of regular cola	5.2	14.7	**32.1**	20.7	11.6	6.4	6.5	2.7
125 g pot of low fat yoghurt	33.7	**42.3**	16.8	3.2	1.1	0.4	0.2	2.4
A serving (30 g) of cornflakes (with 125 ml semi-skimmed milk)	16.6	30.7	**33.0**	12.9	2.8	1.1	0.4	2.6
A serving (small bag) of potato crisps[3]								
25 g/34.5 g	2.3	8.4	**32.0**	22.1	16.0	9.2	7.6	2.3
45 g/50 g	1.1	6.1	18.7	**31.8**	18.8	11.5	8.9	3.1
A bar of milk chocolate (50 g)	1.8	10.0	22.9	**26.9**	17.1	10.2	8.1	2.9
A small glass of wine (120 ml)	8.6	**26.7**	29.2	17.9	8.3	4.2	2.0	3.2
A serving of carrots (80 g)	**66.8**	21.8	5.0	2.2	0.9	0.3	0.3	2.7
A 330 ml can of beer[4]	2.0	8.7	**21.5**	25.5	18.9	11.8	8.6	3.0

[1] Carp in Hungary, cod in other countries.
[2] Wels in Hungary, smoked salmon in other countries.
[3] 25 g for Sweden and Hungary; 34.5 g for the UK; 45 g for France; 50 g for Germany.
[4] In the UK: a pint of beer.

Most of the almost 6,000 respondents in the study knew that health experts recommend to eat a lot of fruit and vegetables (97%). However, only 15% thought that a lot of starchy food such as bread, rice, pasta or potatoes, should be eaten, a recommendation that indeed has been debated and sometimes modified recently. This ranged from 33% in Germany to less than 5% in the UK and Hungary. Generally, respondents tended to exaggerate their response towards the questions about health expert recommendations on foods and drinks high in fat, or in sugars, or in salt, with most of them believing they should try to avoid eating them, rather than consuming a little. The understanding about the consumption of fats is quite mixed, with some confusion about fat quality. Most (62%) correctly thought one should consume more omega 3 fatty acids, and avoid or consume less saturated fat (75%). Over 25% of the respondents did not understand what trans fatty acids were, but most correctly proposed one should try to eat less or avoid them. There was some confusion over how much polyunsaturated fatty acids health experts recommend to eat; only 25% correctly answered more, and over 50% of respondents thought one should try to avoid or eat less, or didn't know what polyunsaturated fatty acids means. The majority of respondents correctly stated that health experts recommend consuming less salt, or less sodium, but 14% did not understand what sodium means.

When presented with a list of 18 different foods or drinks and asked to indicate whether each was high or low in fat, sugar, saturated fat or salt, on average UK, German and Hungarian respondents correctly answered about 70% of the questions, Swedish and French respondents 60%, and Polish respondents 57%. Over 80% gave correct answers for chocolate high in sugar. However, a majority of respondents (60%) incorrectly thought margarine was high in saturated fat and many respondents (47%) thought red meat was low in saturated fat, rather than high (38%).

When asked how many calories are provided by a stated portion-size amount of each of 8 products, UK respondents were most knowledgeable, with about 50% of questions answered correctly; in the other countries it was between 30 and 40%. Knowledge about calories in a portion of carrots was best, with 67% correct answers. When answered incorrectly, the majority tended to over-estimate the calorie content of all foods, except cornflakes and yoghurt, where they underestimate the calories. On the other hand, the calorie content of alcoholic drinks (wine and beer) was over-estimated by respondents in all 6 countries.

Thus, the overall level of knowledge on nutrition recommendations was relatively good in all countries with a majority of respondents knowing which foods should be favoured or eaten occasionally, according to recommendations. The knowledge on nutrient sources was also good for main macronutrients and food-related recommendations, but was less good with regard to details on fat quality (e.g. monounsaturated fat, trans-fat and even saturated fat) and with regard to sodium compared to salt. The results indicate that consumers are aware of the recommendations, but the nutritional knowledge on more technical and detailed issues is less strong.

Knowledge on nutrition sources was also relatively good, but people were less able to assess the energy content of foods. This task is more difficult as it requires both knowledge of the energy content (or macronutrient composition) of the food and then taking into account the serving size.

It has been a much-debated finding that the level of nutrition knowledge differs among socio-economic groups. Results suggest that women and people with higher education levels have more nutrition knowledge (De Vriendt *et al.*, 2009; Grunert *et al.*, 2012; Hendrie *et al.*, 2008). Also age has been positively linked to the level of nutrition knowledge (De Vriendt *et al.*, 2009; Grunert *et al.*, 2012; Hendrie *et al.*, 2008; Wardle *et al.*, 2000). The results in Table 2.1, when broken up according to countries, also suggests that there are differences between countries, with the UK having highest knowledge scores on most types of knowledge. In France the knowledge about trans fats was the lowest and high sources of salt were less known in France and Sweden, whereas knowledge on sugar was high in Hungary and Poland. These differences are probably related to the main focus of the health campaigns that have been carried out in these countries. The history of nutrition and public health policies, along with the history of industry and retailing initiatives in positioning products based on their nutritional properties, together with cultural differences, are a major factor affecting how people acquire knowledge about food and health.

2.3 How do consumers choose their food?

As the previous section showed, many consumers actually know a good deal about the healthfulness of different food stuffs. This does not necessarily imply, though, that they will apply this knowledge when choosing which foods to buy.

Many food choices are habitual, and of those that are not, most are made at the point of purchase. Figure 2.1 shows, for a range of products and countries, the average decision-time that people use – that is, the time from arriving at the shelf to the time when leaving again with at least one product in the shopping cart (adapted from Grunert *et al.*, 2010). As can be seen, the average is around 25-30 seconds, and the distribution around the average is actually skewed, with many people only using 5-15 seconds and a few people using several minutes. In other words, decisions in the supermarket are made very quickly and will not involve a lot of deliberation. Research using eye-tracking methodology has also shown that

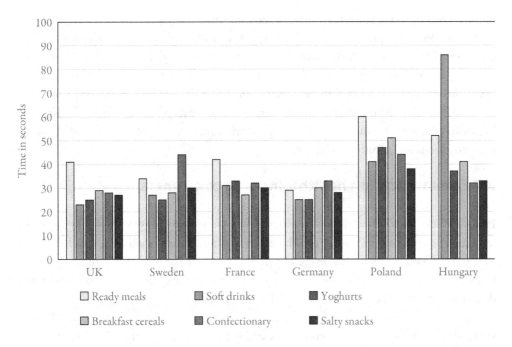

Figure 2.1. Decision times when shopping for food (based on data from Grunert *et al.*, 2010).

when shoppers scan products on the shelf, most products only get a few seconds or less in the process (e.g. Königstorfer and Gröppel-Klein, 2012).

This means that, for many shoppers and in many cases, the decision-making in the store will not be of the kind where the shopper processes several product attributes, like the ingredients list, the nutrition label, the origin and the appearance, and tries to aggregate them into an overall evaluation before making a decision. Most food purchases will not involve a lot deliberation. In dual processing theories of decision-making, one distinguishes between conscious, deliberate decisions that involve reasoning, and spontaneous, intuitive decisions with limited conscious cognitive activity (Evans, 2008). Nobel prize winner Kahneman has labelled these two types of decisions as governed by system 2 (the conscious, deliberate system) and system 1 (the spontaneous, intuitive decision) (Kahneman, 2011). Most food purchases are most likely governed by system 1.

System-1 governed decisions are triggered by a few environmental cues. Shoppers may recognize the brand as one they have used before and found satisfactory. Or they have not used the brand before, but recognize it as familiar and have some positive associations with it, possibly due to being exposed to market communication. They may like the appearance of the project or its packaging. Health-related information will not play a big role in system 1 governed decisions, unless the healthfulness of the product can be linked to those few cues that govern the decision. A product that was previously categorized as healthful may be recognized and bought again. Also, health-conscious consumers may process key cues that they think are related to healthfulness, for example that the product is organic (Thøgersen *et al.,* 2012) or that the product carries a health symbol like the Choices logo or the Keyhole logo. But relatively few shoppers turn the product around to read the list of ingredients or the nutrition table (Grunert *et al.,* 2010).

2.4 Communicating healthfulness to consumers

Healthfulness of a food product is a credence characteristic (Fernqvist and Ekelund, 2014). It cannot be seen, felt or tasted, and therefore any food product which is developed and positioned based on its health properties needs to master the task of communicating the healthfulness of the product to consumers. Given what has been said about consumer decision-making in the previous section, this is a difficult task.

Fortunately, healthfulness can be communicated in many ways. First, there is objective health information in the form of nutrition claims like 'low fat', health claims like 'reduces the risk of coronary heart disease', and various forms of nutrition labelling that communicate the product's content of calories and key nutrients, like fat, sugar and salt. Manufacturers can communicate process characteristics of the product, like being organic or locally produced, which are not directly related to health but which are known to be taken as indicators of healthfulness by some consumers. Manufacturers can use imagery in their advertising and on the food label that is bound to prime the health concept in the mind of consumers, like using pictures of healthy people, people exercising, pure water or serene unspoilt landscapes. Manufacturers can try to have the whole brand built up around a health theme and have that mirrored in the choice of brand names, like in the case of Tesco's 'Healthy Choice' brand. Figure 2.2 shows an advertisement for a dairy project with a health positioning, which combines several of these instruments: the imagery conveys a healthy family with a pure water background, the ad contains a health claim, and it also communicates that the product is organically produced.

Figure 2.3 shows results from a survey asking consumers which indicators of healthfulness of a food product they find important. Findings indicate that while consumers indeed regard the content of nutrients as important, they also regard a low degree of processing, the absence of additives, or the absence of GMO material as indicators of healthfulness, even though the actual link of such indicators to the objective healthfulness of a product is doubtful. A study by Chrysochou and Grunert (2014) confirms that the objectively most

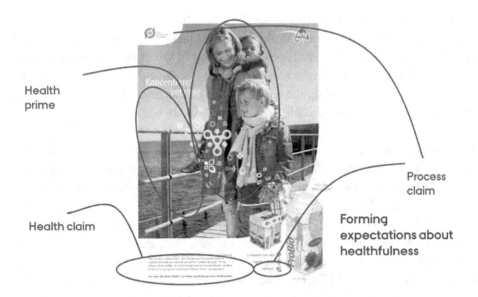

Figure 2.2. Example of ad communicating healthfulness.

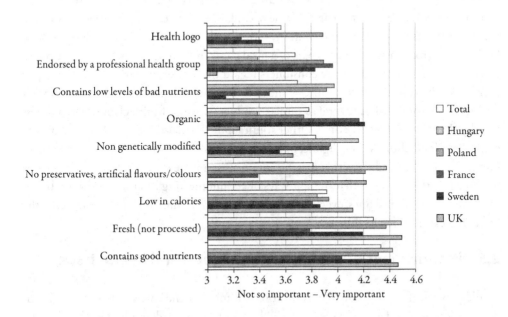

Figure 2.3. Perceived importance of different indicators for the healthfulness of a food product (based on data from Grunert *et al.*, 2010).

relevant indicators of healthfulness are not those that have most impact on consumers' perception of healthfulness. In their study, they tested several advertisements that meant to convey healthfulness by different types of imagery, by health or nutrition claims, and by process claims like that the product was organically produced, or that the product was local or by a small manufacturer. They found that the health and nutrition claims had least impact on the perception of overall healthfulness, while the imagery had most impact.

On this background, it is not surprising that the role of health claims in the marketing of healthy food products has been ambivalent. Many actors in the food industry had hoped that food with improved health properties advertised by health claims would be a major route to growth and new value-added products. Many have therefore been concerned about the regulation of health claims. In the EU, health claims are regulated by the EU Commission, which in turn bases its decisions on recommendations from the European Food Safety Authority, which compiles scientific dossiers on all health claims that are filed for approval. It has been argued that the EFSA-EU system is unnecessarily restrictive, and that this is a barrier to growth for the food industry. However, research has shown that many consumers are rather sceptical regarding health claims, especially when they are on products where the additional health properties are obtained by enrichment or other forms of product modification. Such products are easily perceived as lacking naturalness, which in the mind of many consumers is very close to healthfulness. And when the health properties of products are communicated in the form of technically formulated health claims, like 'Consumption of beta-glucans from oats or barley as part of a meal contributes to the reduction of the blood glucose rise after that meal', many consumers regard this as not attractive. As an example, a major study in the Nordic countries showed that most consumer find products with health claims not only less natural as similar products without health claims, but even less healthy (Lähteenmäki *et al.*, 2010). Other studies indicate that the role of health claims in consumers' food choices is negligible, unless consumers have specific health concerns and find a product that addresses this specific concern (or a review see Wills *et al.*, 2012). For the broad market, it seems that the soft ways of communicating healthfulness work better than the more technical ones, even though these may, from a nutritional perspective, by the better indicators of healthfulness.

2.5 Different ways of dealing with the food and health issue

While choices of food products in the store will most of the times not involve a lot of deliberation, as noted above, consumers still have to face the trade-off between health and pleasure that results from the omnipresent pressure towards healthy eating on the one side and the desire to buy, prepare and eat food that is tasty, pleasurable and liked by the family on the other side. Different people handle this trade-off in different ways. Chrysochou *et al.*

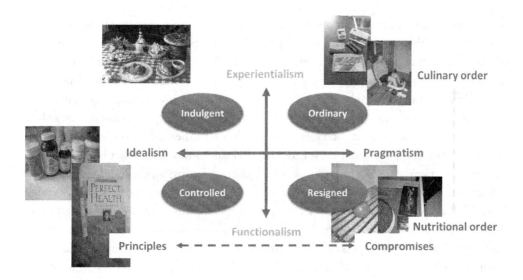

Figure 2.4. Four discourses on food and health (adapted from Chrysochou *et al.*, 2010).

(2010) have analysed the way in which people handle these trade-offs by distinguishing four different discourses that characterize the way people think about the food and health issue. The four discourses come about by combining two dimensions, as shown in Figure 2.4. The first dimension distinguishes between experientialism and functionalism. The experiential discourse represents consumers' view on food within the culinary context, incorporating the notions of gastronomy such as pleasure, taste and quality. The functional discourse represents consumers' view on food within the context of nutritionism that is associated to a bio-medical notion of health and where food is viewed as a bundle of nutrients. Consumers tending towards an experiential discourse may sacrifice healthy eating for the sake of experiencing quality and taste in their food choices, whereas consumers tending towards a functional discourse may – to achieve physical health – compromise on taste or quality by optimising the nutritional content of their food choices. In real life, however, consumers are able to construct a variety of compromise positions. Therefore, the second opposition relates to the idealistic discourse as opposed to the pragmatic discourse towards principles of the culinary and/or the nutritionist order of eating. Idealism can be linked to discourses expressing strong and firm principles concerning culinary and/or nutritionist eating ideals. In contrast, pragmatism can be linked to discourses prescribing a striving to attain guilt-free compromises between healthy eating ideals and competing life demands and situational circumstances. Therefore, consumers representing the idealistic discourse have firm principles towards healthy eating, whereas consumers expressing a pragmatic discourse have a moderated view of healthy eating and more often set compromises.

The two oppositions define four discursive subject positions that are named 'the ordinary', 'the indulgent', 'the controlled' and 'the resigned'. A subject position is one possible outcome of the power struggle between various discourses. The ordinary position covers a look upon eating with focus on the experiential dimension in comparison to the functional dimension of eating. Therefore, they give more importance to notions of gastronomy such as pleasure, taste and quality. This position is generally not very sympathetic towards a strict dietary regime since it intervenes too much with inherited culinary principles of everyday lifestyle. As a result, there is little trust in health information that is based on the notion of a strict dietary regime. Similar to the ordinary position, the indulgent position is focusing on the experiential discourse compared to the functional discourse of eating. This subject stands out through the firm principles towards culinary pleasure and the notion of good life that are expressed by an idealistic approach towards food consumption. The controlled position expresses a prioritization of the functional discourse of eating in comparison to the experiential discourse. It is furthermore characterized by an idealistic approach, thus stressing firm principles towards healthy eating. Eating principles promoted through this position are thus health conscious and concerned about the nutritional content of food. As a result, constant information seeking about healthy eating is encouraged. The resigned position represents a prioritization of the functional discourse of eating in comparison to the experiential discourse. To a certain degree, opinions aligned with the resigned position resemble those aligned with the ordinary subject position with a pragmatic alternation between principles and indulgence. This imbalance is explained as a result of external conditions and constraints (e.g. time, money, availability), and is partly caused by a weak personality or lack of personal discipline. From this position some knowledge is acknowledged concerning what is healthy or unhealthy, but consumers here often describe a feeling of being overruled by one's own desires for the 'forbidden' and lacking the necessary discipline to follow healthy eating.

Every individual consumer is positioned somewhere in the space defined by those four positions, and their positioning will have an impact on how this person reacts to information about healthy eating and about the healthiness of foods, how s/he goes about planning their meals, how they compose meals from ingredients, how they prioritize different food product categories, and not least how they react to new products coming on the market that have a health positioning. For example, it has been shown that consumers scoring high on the indulgence position have a higher tendency to adopt food supplements, possibly because of the promise to achieve health without compromising on the food eaten (Aschemann-Witzel and Grunert, 2015), whereas only consumers with a high score on the controlled position are interested in adopting salt-reduced food products.

2.6 The potential for healthy food products

Food products with a health positioning have been a growth area, but it is by no means easy to profit from the health trend as a food manufacturer, and doing so requires a good understanding of the market and of the way consumers deal with the food and health issue.

First, we should note that not all healthful food product innovations coming on the market are consumer-driven. The enormous attention the food and health issue has achieved in the public arena has put pressure on the food industry to develop healthier alternatives, and the proliferation of nutrition labelling has resulted in a much higher transparency regarding the content of key nutrients in food products. Even though most consumers may ignore this information, many food manufacturers have been reluctant to keep products on the market that are obviously very high in their content of fat, sugar and/or salt, and have engaged in product modifications with the aim to give their products a better nutritional profile. It has been argued that this perhaps has been the most tangible effect of the introduction of nutrition labelling (Vyth *et al.*, 2010).

There are lots of opportunities for developing products that are specifically tailored to consumers with a particular health concern. Consumers with a cholesterol problem or that need to deal with osteoporosis or osteoarthritis will be receptive to products addressing that particular issue and will react positively to corresponding health claims, especially when the product otherwise lives up to their demands regarding taste, appearance and convenience. The same will be true for elderly people with, for example, a protein deficit, even though many of them will evaluate protein-enriched products critically in terms of naturalness.

For the broad market of those not dealing with specific health concerns, and especially for those scoring high on the indulgent and ordinary discourses, the success of products marketed for their healthfulness will hinge on two major factors. The first has to do with the limitations of consumer decision-making discussed above. The healthfulness of the product must be communicated in a clear and credible way without having to resort to technical descriptions in the forms of nutrition tables and health claims. In many cases, consumers can recognize healthy alternatives when they see them, even without the use of nutrition label information (Aschemann-Witzel *et al.*, 2013). New food products coming onto the market should therefore be easily perceived as a healthy alternative based on their overall appearance, supported by appropriate market communication, with health and nutrition claims only having a supplementary function. Especially products that can credibly be positioned as mitigating the perceived trade-off between health and indulgence have good prospects. Many actors in the food industry have recognized this and are working on product reformulations that will achieve this goal.

References

Aschemann-Witzel, J. and Grunert, K.G., 2015. Influence of 'soft' versus 'scientific' health information framing and contradictory information on consumers' health inferences and attitudes towards a food supplement. Food Quality and Preference 42: 90-99.

Aschemann-Witzel, J., Grunert, K.G., Van Trijp, H.C., Bialkova, S., Raats, M.M., Hodgkins, C., Wasowicz-Kirylo, G. and Koenigstorfer, J., 2013. Effects of nutrition label format and product assortment on the healthfulness of food choice. Appetite 71: 63-74.

Capacci, S., Mazzocchi, M., Shankar, B., Macias, J.B., Verbeke, W., Pérez-Cueto, F.J., Kozioł-Kozakowska, A., Piórecka, B., Niedzwiedzka, B., D'Addesa, D. and Saba, A., 2012. Policies to promote healthy eating in Europe: a structured review of policies and their effectiveness. Nutrition Reviews 70(3): 188-200.

Chrysochou, P. and Grunert, K.G., 2014. Health-related ad information and health motivation effects on product evaluations. Journal of Business Research 67: 1209-1217.

Chrysochou, P., Askegaard, S., Grunert, K.G. and Kristensen, D.B., 2010. Social discourses of healthy eating. A market segmentation approach. Appetite 55: 288-297.

Dallongeville, J., Marécaux, N., Cottel, D., Bingham, A. and Amouyel, P., 2001. Association between nutrition knowledge and nutritional intake in middle-aged men from northern France. Public Health Nutrition 4: 27-33.

De Vriendt, T., Matthys, C., Verbeke, W., Pynaert, I. and De Henauw, S., 2009. Determinants of nutrition knowledge in young and middle-aged Belgian women and the association with their dietary behaviour. Appetite 52: 788-792.

Evans, J.S.B., 2008. Dual-processing accounts of reasoning, judgment, and social cognition. Annual Review of Psychology 59: 255-278.

Fernqvist, F. and Ekelund, L., 2014. Credence and the effect on consumer liking of food – A review. Food Quality and Preference 32: 340-353.

Grunert, K.G. and Wills, J.M., 2007. A review of European research on consumer response to nutrition information on food labels. Journal of Public Health 15: 385-399.

Grunert, K.G., Fernández-Celemín, L., Wills, J.M., Bonsmann, S.S. and Nureeva, L., 2010. Use and understanding of nutrition information on food labels in six European countries. Journal of Public Health 18: 261-277.

Grunert, K.G., Wills, J., Celemín, L.F., Lähteenmäki, L., Scholderer, J. and Bonsmann, S., 2012. Socio-demographic and attitudinal determinants of nutrition knowledge of food shoppers in six European countries. Food Quality and Preference 26: 166-177.

Handu, D.J., Monty, C.E. and Chmel, L.M., 2008. Nutrition education improved nutrition knowledge, behavior, and intention among youth in Chicago public schools. Journal of the American Dietetic Association 108: A91-A91.

Hendrie, G.A., Coveney, J. and Cox, D., 2008. Exploring nutrition knowledge and the demographic variation in knowledge levels in an Australian community sample. Public Health Nutrition 12: 1365-1371.

Kahneman, D., 2011. Thinking, fast and slow. Macmillan Publishers, London, UK.

Klohe-Lehman, D.M., Freeland-Graves, J., Anderson, E.R., McDowell, T., Clarke, K.K., Hanss-Nuss, H., Cai, G., Puri, D. and Milani, T.J., 2006. Nutrition knowledge is associated with greater weight loss in obese and overweight low-income mothers. Journal of the American Dietetic Association 106: 65-75.

Königstorfer, J. and Gröppel-Klein, A., 2012. Wahrnehmungs-und Kaufverhaltenswirkungen von Nährwertkennzeichen auf Lebensmitteln. Marketing ZFP 34: 213-226.

Lähteenmäki, L., Lampila, P., Grunert, K., Boztug, Y., Ueland, Ø., Åström, A. and Martinsdóttir, E., 2010. Impact of health-related claims on the perception of other product attributes. Food Policy 35: 230-239.

Lee, J.W., Lee, H.S., Chang, N. and Kim, J.M., 2009. The relationship between nutrition knowledge scores and dietary behavior, dietary intakes and anthropometric parameters among primary school children participating in a nutrition education program. Korean Journal of Nutrition 42: 338-349.

Parmenter, K. and Wardle, J., 1999. Development of a general nutrition knowledge questionnaire for adults. European Journal of Clinical Nutrition 53: 298-308.

Sapp, S. and Jensen, H.H., 1997. Reliability and validity of nutrition knowledge and diet-health awareness tests developed from the 1989-1991 Diet and Health Knowledge Surveys. Journal of Nutrition Education 29: 63-72.

Thøgersen, J., 2009. Consumer decision-making with regard to organic food products. In: De Noronha Vaz, T. and Nijkamp, P. (eds.) Traditional food production and rural sustainable development: a European challenge. Routledge, London, UK, pp. 173-192.

Thøgersen, J., Jørgensen, A.K. and Sandager, S., 2012. Consumer decision making regarding a 'green' everyday product. Psychology and Marketing 29(4): 187-197.

Vyth, E.L., Steenhuis, I.H., Roodenburg, A.J., Brug, J. and Seidell, J.C., 2010. Front-of-pack nutrition label stimulates healthier product development: a quantitative analysis. International Journal of Behavioral Nutrition and Physical Activity 7: 1-7.

Wansink, B. and Sobal, J., 2007. Mindless eating the 200 daily food decisions we overlook. Environment and Behavior 39: 106-123.

Wardle, J., Parmenter, K. and Waller, J., 2000. Nutrition knowledge and food intake. Appetite 34: 269-275.

Wills, J.M., Bonsmann, S.S., Kolka, M. and Grunert, K.G., 2012. European consumers and health claims: attitudes, understanding and purchasing behaviour. Proceedings of the Nutrition Society 71: 229-236.

3. The sustainability trend

D. Brohm* and N. Domurath

INTEGAR GmbH, Schlüterstr. 29, 01277 Dresden, Germany; brohm@integar.de

Abstract

A lot of people in the Western World wonder about a lot of things. Especially things close to their personal environment. Food is a very private matter, because we eat and drink every day and absorb both healthy and maybe harmful components from the food. A range of incidents, scientific studies and ethical considerations of the past decades has led to a major interest of the public in where our food comes from and how it is treated and processed. Since the implementation of chemical plant protection and fertilisation in agriculture in the beginning of the 20th century, a small group of people rebelled against the new possibilities. They preserved traditional forms of plant cultivation and animal breeding and developed what we now know as organic production. Another development of the past century is Fair Trade. The exploitation of men and resources in the Third World came to the attention of the public in the early nineties by television. Especially child labour became a main topic. Again a minority of consumers began to complain and started first projects to fight the tort and improve the situation of the farmers. Since the whole world is connected by internet, the number of people that are interested in these problems increased dramatically and new topics regarding sustainable development of food production came up. So also the question of origin came into focus. In the beginning may be it was important to buy food from the home country. Nowadays, the real distance between production area and point-of-sale is more important to the consumers. Regional food, local food and also self-supply are issues that rise more and more especially in urban areas. A current topic is food waste. Here the concept of Food Sharing is an answer from the people.

Keywords: organic, origin, CSR, Fair trade, food waste

3.1 Organic production

With the beginning of the technical and chemical intensification in the agricultural sector in the end of the 19th century, the loss of the peasant traditions began. The whole farming economy and ecology changed within a short time. New solutions also generated new problems, like decrease of soil fertility and increase of pests and diseases caused by monoculture, or bankrupts of traditional farms that could not adapt to the new developments. Movements against urbanisation and industrialisation wanted a return to a natural way of living with aspects like vegetarianism, naturopathy, physical culture, self-supply in allotments and garden cities, nature conservation and animal welfare (Vogt, 2001). Some of these keywords are still up to date and even more in the focus of the public. During the 20th century, a couple of different organic cropping systems were developed. A legal definition of organic production came up in 1990 with the 'Organic Foods Production Act' in the United States and 1991 with the first EU-Eco-Regulation. So far, 69 states have implemented legal standards for organic farming, and 21 other states are working on it. All these laws regulate, amongst others, the use of fertilisers, pest treatments, medication of animals, but also soil management, breeding and marking of organic products. For some of the farmers and consumers these regulations were not strict enough. Organisations, mostly based on existing cropping systems and/or particular attitudes, developed stricter guidelines and new labels by their own. Demeter, for instance, is the oldest certification organisation, founded in 1924. Its philosophy is predicated on biodynamic agriculture that is closely connected to the anthroposophical ethos. Others are Bioland, Naturland, Gäa and Ecovin.

A couple of incidents during the past decades led to a major loss of trust in the food industry, but also to more interest of the public in where our food comes from and how it is treated and processed. Examples of serious events are the outbreak of *Escherichia coli* bacteria (EHEC) or the use of horse meat as beef. But also ethical considerations in relation to animal husbandry or genetically modified organism have occupied a lot of people. Finally, scientific studies based on new measuring methods are showing chemical residues of plant protection chemicals in fruits and vegetables. Organic food evolved from a niche to a trend. Supermarkets only for organic products came up and almost every food discounter chain developed product lines only with organic products. All these developments were promoted by the upcoming media of the internet.

Over the past decade, the demand for organic products and the organic agricultural area have increased significantly, as Figure 3.1 shows (Willer and Lernoud, 2016). The boom was so great that even discounters have developed their own product lines and labels for organic foods. This soon resulted in a great pressure on the producers. The quality must be high and the prices as low as possible. At the same time, productivity in organic farming is much

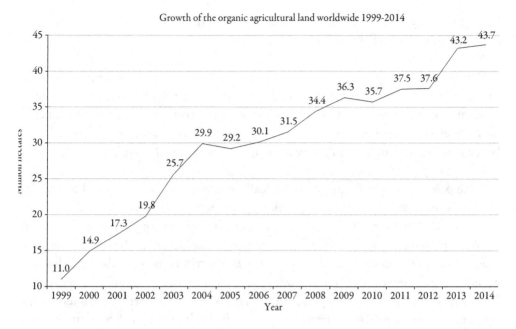

Figure 3.1. Growth of the organic agricultural land worldwide 1999-2014 (Willer and Lernoud, 2016).

lower. On the other hand, discounters also set conventional producers under pressure by restricting the permitted maximum legal residue of biocides or by totally forbidding some of them (ISN News Desk, 2016). One consequence is that both forms of cultivation more and more approach each other. This in turn exerts pressure on the organic producers. Many farmers who have converted to organic production reversed their decision. The additional expenses in organic farming were financially not manageable, since the prices of these products have decreased further and further.

But how did this organic boom happen? The increasing demand for organic food came from consumers. A couple of reasons are responsible for this development. Many people care about what they eat. In an affluent society, where everything is available at any time, everyone can decide freely how to feed. Even processing and cultivation form is integrated into the selection. The modern organic consumer lives in a city, is well educated, has a pronounced environmental consciousness and is self-determined (Dettmann and Dimitri, 2009). Sustainable action is important to him to secure his descendants a healthy world. But living in a city also reduces the relation to nature. The idea of how foods are made is often incompatible with reality. Nevertheless, the consumption of organic food evokes a greater sense of having done something good for the community and the personal health.

Because of many issues in agricultural production and the food industry during the past, consumer trust has eroded.

A high demand for organic food has been limited to the developed countries so far, even though a lot of organic products are already produced in developing countries. The main disadvantage of this form of cultivation is its low productivity compared to conventional forms (Seufert *et al.*, 2012). Calculations show that the increasing world population can only be fed in the long-term by an intensive agriculture (Godfray *et al.*, 2010).

On the one hand organic products basically contain less pollution of biocides or medicaments. They are also freer of genetically modified organisms. On the other hand it is not clarified if they are more healthy compared to conventionally produced food or even if conventional food is basically less healthy. Also, potentially higher health benefits of organic food because of higher contents of vitamins, mineral nutrients and secondary metabolisms are not comparable with common eating habits, because the prevailing mixed diet does not lead to symptoms of deficiency (Strassner, 2011). An analysis of over 200 studies on this subject found no significant differences between the health benefits of conventional and organic food (Smith-Spangler *et al.*, 2012).

There's no accounting for taste. If organic products actually have a better taste has not yet been analysed comprehensively. The main reason is the ability of the human sense of taste to adapt to many things. People with a solely organic diet have a different sense food from conventional production and vice versa. Also a scientific inspection proves difficult. In fact, instruments can analyse flavouring ingredients. But the overall taste cannot be realised without sensory examinations (Strassner, 2011).

As long as food security is guaranteed, alternative production systems with extended requirements will exist side by side with common established ways of food production. These requirements can be based on health, environmental protection, ethics or spirituality. Currently, more efficient production methods are in a developing process. Those methods exceed existing food production systems in aspects of productivity, reduction of land consumption and resource conservation. An unprejudiced elucidation about the available systems could point out the existing advantages and disadvantages of each method. The more food demand rises on earth, the more organic food production will be an alternative way for well-established nations and persons. In the end consumers decide if organic production will sustain its position.

3.2 Fair trade

In addition to the health benefits of a product, for many people the circumstances of how a product has been produced became more important. Especially the equitable commerce with the producers is in focus. The first trade associations for equal trade were founded in the 1960s as an alternative to business models of big enterprises with the aim to exploit human resources in foreign countries mainly relating to handcrafted goods. During the 1970s the attention of fair trade turned to agricultural raw materials. Especially sugar, coffee and cocoa were available as fair trade goods in increasing quantities. Until the 1990s, the sale to the end customer was carried out almost exclusively by so-called world shops. Over the years, more and more goods were added to the range of products, in particular processed goods like chocolate and other sweets. By establishing trusted labels such as 'fair trade', a number of products made it into the supermarket shelves. Besides raw materials and processed goods, nowadays also fruits, vegetables, ornamental crops, flowers and even cotton for the textile industry are available.

But the requirements for fair trade products have evolved over the years. Previously it was mainly about fair remuneration. Now the protection of the environment and natural resources are also taken into consideration as well as the adherence to social standards like equal status of sexes or the refusal of child labour. Lately also goods of modern life are available. One example is the so called Fairphone. This smartphone is made up of mainly fair traded components and raw materials (TransFair, 2016). During the past years, a couple of different labels for different issues came up. Well-known labels are dealing with CO_2-neutrality, child labour, sustainability in the area of agriculture, forestry and fishing or social and environmental responsibility.

Since adding more product groups and aspects to the fair trade concept, the organisation became more visible. But a boom like it happened to 'organic food' did not happen to the fair trade idea so far. A reason for this could be the low level of personal connection of the consumers to the actual manufacturing process and the entire circumstances of the production. But it is also possible that fair trade remains a nucleus for further aspects which later can be transferred into national or international standards.

3.3 Origin

We are living in a globalised world. The origin of agricultural products plays a major role for costumers. An example is apples from New Zealand. They are available in European super markets for the same price as domestic apples. The same applies to other products and foods. Informed consumers know that many goods are produced cheaper in other countries. The low price in turn increases the pressure on the domestic growers. The labelling of the origin of agricultural products has been a legal requirement for years. Other products are marked in order to highlight the origin. In some developed nations more and more customers prefer domestic food products instead of imported goods. Their reasons are strengthening of the internal market, better quality, shorter ways of transportation, less CO_2 emission, better customer service and higher identification with the product based on traditions. This behaviour is especially pronounced in agricultural products. Domestic foods in the broadest sense are in more demand than if they originate from farther away. Exceptions are products that cannot be grown at home for climatic reasons (e.g. tropical fruits) or products that can be grown only in a certain season (e.g. asparagus and strawberries in the winter time). But the distances between the places of food production and the point-of-sale became increasingly shorter over the years. In particular, many consumers in urban areas want agricultural products that are grown nearby. Nowadays foods are labelled as domestic, regional or local. Super markets adjusted to this and tried to decentralise their suppliers in order to be able to promote especially fresh food with 'from the region'. Many local small businesses benefit from the ongoing trend. They sell their goods mainly at farmers' markets. The prospective success of locally grown products also depends on consumers, as long as the origin of food is not regulated by law, because prices play a crucial role. Even if local products are popular, a lot of people cannot afford them and decide for low priced alternatives that could come from far away. Examples are asparagus from Peru or apples from New Zealand.

One step further are people who want to realise self-supply. 100 years ago, this was possible without difficulty. But nowadays it is very hard, because the society is based on division of labour. Nevertheless a lot of people want to produce fresh food, mainly fruits and vegetables, themselves. Balconies, terraces and even indoor facilities are used to grow plants. One decade ago allotments seemed to be old-fashioned recreational activities for seniors. But today there are long waitlists. Especially young families with kids are using this opportunity to teach their children where food comes from. But actually a city is not the best place to realise an all-encompassing food supply. Not for one person and not for all. But there are first projections and preparations to produce particular foods under artificial conditions within urban areas. But the acceptance of the consumers for this type of local food production is just as important as the economic feasibility.

3.4 Food waste

Another current trend in the sustainability movement is directed against the waste of edible food. Approximately 1.3 billion tons of foods are wasted every year. On the other side, almost 800,000 people are starving. Various reasons are responsible for this. A lot of foods are destroyed by manufacturers and especially supermarkets, even though they were probably still edible. For many years, there have been organisations (e.g. food banks) that collect these goods and distribute them freely to people in need. Nevertheless, it is not in the interest of supermarkets to give away food for free. So some governments have already passed laws prohibiting the disposal of edible food. Also, the concept of best-before-date on foods is up for discussion in some states. Another reason is the change in the size of households. Today more single-person and childless households exist than 20 years ago. But here the food industry already reacted by offering smaller packaging.

But also private persons are able to give away food for free that is not needed anymore. In many cities people can share food with other persons. Refrigerators in central places keep the food fresh and everybody can put something in or take something out. The organisation and communication of 'foodsharing' projects use to work by public social platforms. So it is also possible to get free food anonymously.

But the best way for consumers to prevent food waste is to avoid it from the beginning. So it is better to buy smaller portions. Also planned shopping helps to only buy food that is required for the next days. A lot of food gets wasted directly in the supermarkets, because of small blemishes on fruits and vegetables. This is especially absurd in cases of foods that are processed again at home. All food has to be stored in its best way and leftovers can be frozen for later meals.

In the European Union food is very cheap compared to other goods and the average appreciation for food is not very high. Almost every product can be imported from outside and customs duties are low for foods. A solution for the problem of food waste can only to be found, if all parties, producers, resellers, costumers and governments are working together.

3.5 Corporate social responsibility

The demand for products that can be purchased without a guilty conscience continued to rise in recent years, as described. An expansion to almost all areas has taken place. Meanwhile, in many nations standards for socially and environmentally sustainable action and production are regulated by law. In addition, there are numerous companies that commit under the term 'corporate social responsibility' (CSR) to sustainable action beyond the statutory requirements. However, this concerns not only the goods and services of a company, but often the entire field of responsibility, including the treatment of employees and affiliates, e.g. the sub-suppliers. On the one hand the higher costs and efforts are in conflict with the economic principles. On the other hand it ensures the establishment of integrity and credibility, without an acceptance of which the society and other companies would be jeopardised. Compliance with the standards is made credible by granting certificates.

A famous example for good CSR in the food industry is the ice cream manufacture Ben & Jerry's. Its founders have infused the company with the notions of giving back in every way possible, as well as 'linked prosperity' between the company, its employees and the community. They started the Ben & Jerry's Foundation, were founding members of the Business for Social Responsibility organisation and set an extraordinary rate of giving to charitable organisations in the corporate world, donating a full 7.5% of pretax profits. In their own words, they 'strive to show a deep respect for human beings inside and outside our company and for the communities in which they live.' Unilever bought Ben & Jerry's in 2000 and continues to support the foundation. (Liodice, 2010)

Also the coffee brewer Starbucks is a good example. Since the company started in 1971, it has focused on acting responsibly and ethically. One of Starbucks' main focuses is the sustainable production of green coffee. With this in mind, it created C.A.F.E. (coffee and farmer equity) Practices, a set of guidelines to achieve product quality, economic accountability, social responsibility and environmental leadership. The company supports products such as Ethos Water, which brings clean water to more than 1 billion people who do not have access (Liodice, 2010).

Both examples show that CSR can be a company policy from the beginning. But CSR also is important for bigger companies, because they are in special focus. The initiative not only comes from the company, it also is promoted by the employees, by costumers, clients and competitors. In any case it is important to publish the efforts of CSR to make them visible. In the future the social responsibility will be a fundamental aspect of many companies if it is not already the case.

References

Dettmann, R.L. and Dimitri, C., 2009. Who's buying organic vegetables? Demographic characteristics of U.S. consumers. Journal of Food Products Marketing 1: 79-91.

Godfray, H., Charles, J., Beddington, J.R., Crute, I.R., Haddad, L., Lawrence, D., Muir, J.F., Pretty, J., Robinson, S., Thomas, S.M. and Toulmin, C., 2010. Food security: the challenge of feeding 9 billion people. Science 327: 812-818.

ISN News Desk, 2016. Aldi asks suppliers to stop using pesticides. Available at: http://tinyurl.com/yaqah3l8.

Liodice, B., 2010. 10 companies with social responsibility at the core. Available at: http://tinyurl.com/9jrsq9f.

Seufert, V., Ramankutty, N. and Foley, J.A., 2012. Comparing the yields of organic and conventional agriculture. Nature 485: 229-232.

Smith-Spangler, C., Brandeau, M.L., Hunter, G.E., Bavinger, J.C., Pearson, M., Eschbach, P.J., Sundaram, V., Liu, H., Schirmer, P., Stave, C., Olkin, I. and Bravata, D.M., 2012. Are organic foods safer or healthier than conventional alternatives? A systematic review. Annals of Internal Medicine 157: 348-366.

Strassner, C., 2011. Besser, gesünder, geschmackvoller? Haben Ökolebensmittel wertvollere Inhaltsstoffe als konventionelle Produkte? Ökologie und Landbau 157: 16-18.

TransFair, 2016. Fairphone 2 mit fairtrade-gold. Available at: http://tinyurl.com/y9v2vgcq.

Vogt, G., 2001. Geschichte des ökologischen Landbaus im deutschsprachigen Raum – Teil I. Erste Ansätze zu ökologischer Landbewirtschaftung gab es bereits Ende des 19. Jahrhunderts. Ökologie und Landbau 118: 47-49.

Willer, H. and Lernoud, J., 2016. The world of organic agriculture – Statistics and emerging trends 2016. Research Institute of Organic Agriculture, Frick, Switzerland, 48 pp.

4. The authenticity trend

M. Petz and R. Haas[*]

Institute of Marketing & Innovation, Department of Economic and Social Sciences, University of Natural Resources and Life Sciences, Vienna (BOKU), Feistmantelstr. 4, 1180 Vienna, Austria; rainer.haas@boku.ac.at

Abstract

A general overview of authenticity in connection with food is given. Cases are presented by way of example to explore authenticity in relationship to modernism, hipster authenticity and hyperrealism. Indicators of authenticity are given that allow a matrix of authenticity to be scored on different metrics. Levels of authenticity are proposed with a scale from A1 (most authentic) to A2 and A3 (least authentic) analogous to the L1-L2-L3 concept in linguistics. Different perceptions of authenticity are discussed, also in relation to origin-linked foods, living traditions, craft authenticity and true authenticity. Authenticity with respect to both manufactured consumer products and home cooked foods is considered. True authenticity is ascribed based on the direct authenticity of a food's characteristics and its ecological authenticity, which considers the cultural milieu and associated aspects of production.

Keywords: cultural omnivouressness, folklorism, nostalgic foods, terroir

4.1 Tracing the source of the authenticity trend

We must dissect what an authenticity trend actually is to discover its source. This can be done by looking at analogous trends that are often conflated or associated with an authenticity trend as a cluster of related phenomena (Johnston and Baumann, 2007). We must also distinguish between authenticity as a marketing hype, true authenticity and authenticity as a trend in cooking, eating or broader consumer patterns. We can identify where we are on the hype cycle of a trend by where we find it and in which publications (Levy, 2015). 'A major role of gourmet food writing is to spot culinary trends and to identify particular dishes and foods as being worthy food choices'(Johnston and Baumann, 2007).

We must be conscious of who says something is a trend; white upper-class consumers, writing in their broadsheet newspapers (Bell and Hollows, 2011; Johnston and Baumann, 2007), food researchers, trade press or company publications such as menus and company reports or even chef's artist manifestos (Adria *et al.,* 2006) can all make some claims to identifying a trend. Looking at the Food Babe Blog (Alsip, 2015) the origins of the posts there can be traced back to an originating product, chef or origin of the innovation, but are likely to be secondary trends. Academic publications like the Journal of Food and Nutrition Research are likely to show trends with a lag and report primary trends after they have begun to trend, even though it claims to have 'rapid publication of articles in all areas of food and nutrition' (Mandal, 2016a) with a 'manuscript accepted ... go[ing] through the peer review process in one months or less'(Mandal, 2016b).

Yet for our purposes we are focused on consumption patterns that may be limited to a certain demographic group defined by, e.g. class, income, generation; or by a shared set of cultural values (Johnston and Baumann, 2007).

Localization and adaptation to local markets may mean that a food is not a truly authentic food, but is merely marketed as such. Exemplifying this, a purported 'Guinness beer', which is a porter stout and a black opaque drink, was served in France at the first author's table in a translucent form closer to a pale ale. A food example is curry. Curry has appeared in the *British Army Cook Book* since the early 19th century (Zlotnick, 1996). Yet over a century later it was still described as a foreign food in the *Friends of The Earth Cookbook* (Sekules, 1980). While the foreign origins of curry would be recognized in the UK, nowadays it would not be seen as a foreign food for anyone aged under 30 and in fact Chicken Tikka Masala has been the most popular dish in the UK for at least 20 years (Mannur, 2009). It could be argued that as a British Empire food, curry was not foreign at all to the UK in the 1930s, thus Friends of the Earth were wrongly categorizing it as such. Instead of merely putting it within the context of 'Indian' cuisine and declaring it as foreign they should have put it as within the cuisine of 'Britishers' and thus, if not a native food of England, at least a familiar part of the indigenous Briton's diet by the 1980s.

However, this is the dish or beverage only. When we look at the wider cultural practices and milieu, we can find true authenticity, 'the real', and also an ersatz mediated authenticity, '*simulacra*' (Munoz *et al.,* 2006), for example with Irish theme pubs and the plastic paddies we find connected to them (Bogle, 1993; Scully, 2009). Here, a 'constructed space' is made by 'reality engineers' that make a 'perceived authenticity' which is a 'hyperrealism' (Munoz *et al.,* 2006).

> engineered Irish pubs are simulations of the real thing and can often be interpreted by consumers as authentically Irish. Yet, differences between the simulation and original abound. ... Themed Irish pubs, for example, are typically more technologically advanced, cleaner, and stylish than their genuine counterparts. ... consumers of Irish pubs are 'insulated' from the imperfections that exist in the 'real' versions. Consumers desire and expect these 'pseudo-events' and marketers provide it; therefore, both are responsible for creating and maintaining the perception of authenticity. This then raises the question 'what is authenticity and how is it perceived?
>
> As Munoz *et al.* (2006) discussed

4.2 Recipes and authenticity

Food culture and food recipes do not stand still, and people often do not recall how food actually was in the past. Research done in South America and Tyrol, Austria on the Austrian diaspora has revealed that the range of traditional foods known as Austrian cuisine is limited and mediated by local food ingredient availability (Kuhn, 2013; Pirker *et al.,* 2012). The addition of an ingredient may mean that a food stops being authentic, for example flour treatment agents or colorants in bread (Whitley, 2011). Yet an 'authentic' dish may use a substitute (DeMers and Fuss, 1998), as is found with 'Cho Cho Apple Pie' (Benjiw, 2011; DeMers and Fuss, 1998), which contains cho-cho, but no apples! It is authentic. If it is part of a living tradition it can be said to be traditional food. More importantly 'Changing the actual physical materials of food has little to do with eating ethnically' (Douglas, 2014), but rather authenticity here is 'based on the function of the respective ingredient... [in respect to] ... taste... the right consistency ... [and that it] fulfils the same function, without impairing the function of the dish'(Kuhn, 2013). To this should be added the cultural practices that surround food (Douglas, 2014), consumption, preparation and marketing (Walsh, 2015).

When re-enactors hold a feast, they try to follow traditional recipes, but are forced to substitute heirloom varieties of vegetables by modern agri-business versions. Food from medieval times may be cooked with potatoes rather than swedes or turnips. So the food is not truly authentic. It is a reproduction of the closest we can mimic, which represents or re-enacts what we think history was like. This striving for authenticity is found in other food cultural situations, and the level of acceptance of authenticity depends on the knowledge of those participating in the respective food culture.

So why is authentic food desired? Authenticity has come about as a desire to reject industrially processed food and modernism. Modernism in the food industry began in the 1920s and 1930s by implementing Fordism, Taylorism and mass manufacturing processes. While there had been an international food industry before, for example in the tea business, which blended, transported and sold blended teas rather than traditional single origin/single estate products, the production processes were not mechanized nor being promoted to replace existing cultures as inferior to the same extent. Similarly, while manufacturing took this approach, the modernization movement was applied to foods, with the rise of the American meal where hamburgers, fried chicken and pizzas became standardized and slightly later in the 1950s consumed at modernized diners and burger bars. In agriculture, further up the food-chain, the modernization process of mechanization and the move toward monoculture and the 'get big or get out' sentiment manifested themselves (Edmundson and Martusewicz, 2013; Ellis, 2008).

While this modernization movement gained predominance in production it also contained an element of technological determinism. The kind of food that works in machine production is not quite the same as the food that works in hand production. Or more charitably, a uniform carrot is easier to grow, harvest, package and sell than a range of carrots that ripen at different times (Barber, 2014) or bendy bananas that won't pack or fruit that bruises easily in the shop – remove that bloom for nice shiny plums! More than just selecting varieties, whole new foods were created, for example butter was replaced with whale fat and then margarine.

As a counterpoint to the modernization trend there were always traditional food cultures. Here food was to be prepared from scratch from the simple, everyday ingredients that were at hand. Simple means here without much in the way of industrial process, see Johnston and Baumann (2007) for a more in depth discussion of simple and simplicity re authenticity. In a living tradition these would be planted in home/kitchen gardens or bought directly from growers or small wholesalers that supplied market stalls and small outlets. This still occurs for example with the wild berry industry in Finland, where a trader is often a farmer that buys the berries direct from pickers and then delivers them to small supermarkets and sellers in street markets. But in many cases the local links are broken by mass distribution and mass production. But the berries are still berries. They are still authentic berries. The authenticity here is neither a reaction nor a response to modernism; it existed before and persists. Further processing and production practices add a premium to these raw ingredients and this aspect has followed the modernist trend. This modernist trend constantly grew, so now it is the dominant trend and the traditional production methods are in the minority. But both have existed in tandem: mass mediated products, and raw ingredients with traditional production methods.

This presents a conundrum particular to authenticity as a food trend, if we remember what a trend is. A trend is not a living tradition; it is not an established part of the varied food-cultural landscape that is already common. Yet with authenticity we are claiming this is a trend, and perhaps it is a waxing of those traditional practices rather than something wholly new. How do we tell if it is *de novo* or a resurgence? And where do we contextualize it? One old farmer's authentic shepherd's pie maybe a hipster's authentic rural experience, yet a retired soldier's typical canteen food, so to one of them it is a part of the authenticity trend, but not to the other just because they are from a different sub-culture?

An example of this can be seen in an artisanal product that was made in Finland by the first author in conjunction with the Finnish Migrant Berry Pickers Union. Here an old jam recipe was sourced from the 1920s – not a modernist version – and this was produced in a batch. The 'roaring 20s jam' was then contrasted with a 'blueberry chutney', also a traditional practice, but not in Finland, and a more 'modernist jam' recipe with blueberries that made use of modernist food aspects such as pre-pectinized sugar called 'hillosokeri' in its production. Which product was authentic? All three could be said to be inspired by a tradition. The chutney would not be a traditional recipe: chutneys are not made in Finland, so blueberry chutney would not meet this standard and thus it could be argued it is not authentic. The roaring 20s jam could be argued as traditional and authentic. Lastly the modernist jam with blueberries would be accurately described as non-traditional and non-authentic.

Yet in marketing that modernist jam may well be described as 'made within the tradition', which is making jams with blueberries. But it would lack authenticity due to the significant alteration in production method and actual ingredients. Part of the marketing mix relied on packaging that re-enforced the idea of authenticity and another part (some ingredients) confirmed to modernity. 'Marketers provide the means for consumers to learn the symbolic associations of products (through advertisements, movies, and themed environments); however, it is up to the consumer to form these interrelationships for themselves' (Munoz *et al.,* 2006).

The authentic *roaring 20s jam* was sold at a Christmas craft market, through the Pispala Contemporary Art Centre and via word of mouth where information was given about its authenticity. Experiences of authenticity were highlighted by a SecondLife machinima, art exhibitions, product tastings and a discourse that connected the jam to nostalgia for berry-picking and cottage industry production in the Finnish culture. The producers of the jam also could claim authenticity, having come from North European family traditions of foraging berries and making preserves from them.

However, there was a departure from real authenticity as the jars that were used were standardized and bought from a supplier. The same kind of jars could be seen in supermarkets also with a blueberry jam in them. Standardized packaging, serving sizes and printed labels rather than handmade or ersatz hand-written labels as can be seen with, e.g. some of the jams of the Vienna company Staud's GmbH. Here the idea maybe to suggest a cottage industry or grandmother's kitchen theme rather than a production of what are high quality jams in a modern industrial production environment. Staud's wants to emphasize high quality and tradition with its packaging, as also shown by other products in its range which use imagery of highbrow cultural icons from Vienna (Smith, 2014).

Elsewhere true authenticity can be seen with varied jar sizes, real handwritten labels and batch variation in production. Some of this is due to cottage industry production from the maker movement subculture (Dougherty, 2012). Some of it is through a deliberate aesthetic that is cultivated to suggest authenticity. This deliberate choosing is part of the authenticity trend. It can be seen with specialty high class grocers and bakeries and some stores that act more as boutiques than warehouses.

4.3 Terroir and traditional foods as brands?

In the EU there are three labelling schemes (protected designation of origin (PDO), protected geographical indication (PGI) and traditional speciality guaranteed (TSG)) that use the concept of terroir as a quality cue (Grunert and Aachmann 2016). 'The aim of these schemes is, of course, to help consumers and protect traditional production schemes. Whether that works is a different question' (K. Grunert, personal communication, 2016). Terroir is a defined geographical area where a culture has developed, an intellectual or tacit production praxis over a long time, based on interactions between the physical and biological environment, where the socio-technological interactions reveal an originality, confer a typicity and thus generate a reputation for things sourced from there (Vandecandelaere, 2010). The terroir-derived products sourced from that terroir can be called 'origin-linked products' or Spécialités de region.

Origin-linked products may contain a particular ingredient, which must have been sourced from the terroir in question. The schemes may also lead to exclusion of other origin-linked products that are within the production tradition, but do not meet some criteria (Vandecandelaere, 2010). An example are Cornish pasties that do not contain a designated vital ingredient such as carrots. In South Africa some wines are not allowed to be marketed in Europe as they sound like French wines. Here authenticity is not the issue, but consumer confusion is (Stern, 1999). The labelling is effectively being used to brand products for consumer marketing.

Parallel to these trends in the food industry there has been a reflection from the NGO or activist world. The third sector has looked on food politics in a broader way. This has been done most successfully with the International Federation of Organic Agriculture Movements (IFOAM) (IFOAM, 2005) that has concentrated on organic food. Organic is not necessarily authentic food. Similarly, a reaction against modernist trends of GMOs has been the 'Greenpeace True Food Campaign' (Ha, 2008) and the Friends of the Earth's activism for 'Real Food' (Gura, 2001), which is even more strongly echoed by Local Harvest's call for 'real food, real farmers, real community' (Bell *et al.,* 2010). There has been criticism of all of these different positions: that they are polemics, and do not fully deal with the entire food production system; or that individual foods may not meet some criterion/a for authenticity. Some examples are seasonality and food miles in production of an authentic product.

4.4 Modernisation and globalisation as an antithesis to authenticity

In everyday life clear criteria are lacking for what is authentic and who is the arbiter of authenticity. In short, authenticity is delegated to the eye of the consumer, albeit in the face of stringent efforts from 'brand culture' to place its marketing spectacles on the consumer nose (Banet-Weiser, 2012). Authenticity is thus still a social construct (Johnston and Baumann, 2007). This treats 'authenticity' as a brand (Banet-Weiser, 2012). And brands are defined not by the marketer or the producer, but by the consumer or client of a product or service. By extension the brand – in this case authenticity – is not solely the product, but is also the processes around it, the cultural milieu or product ecology (Forlizzi, 2008) it is embedded in. An example of this is an 'authentic' cola. As colas have only been in existence since the 1880s there are no traditional (in the sense that a new tradition can start at any time, but has to be embedded in a clear culture) colas we can find before this time. All other colas afterwards must refer back to that cola or analogous invigorative pick-me ups sold as drinks or quack cures (Pendergrast, 2013).

In trying to modernize Coca-Cola there did emerge a concept of the 'real coke' as opposed to the 'new coke' (Oliver, 2013). That real coke was said to be authentic. Here the cultural place of coke and historical context was important (Gilmore and Pine, 2007; Pendergrast, 2013). Although efforts have been made to replace the cultural framing, for example by Pepsi, Mecca Cola, or more co-operative 'red' leaning colas they have not been successful.

What is important here is that the process of cola production has changed and continues to evolve, for example using robot production methods, computing and packaging. The ingredients are changing, and while the broad formulation is maintained, the subtleties that gos into colas change: 'Diet Coke' was introduced with the aim of capitalizing on the '"brand equity" of the Coca-Cola name [and be a] "line-extension"' (Pendergrast, 2013). Thus, for those not aware of this newness in formulation, it manages to capture an aura of authenticity from Coke, despite the fundamental changes in ingredients from sugar to saccharin. In contrast, even though Tab (a clear cola also made by The Coca-Cola Company) is older, predating diet coke and the clear cola movement, Tab did not/does not gain this brand caché.

Thus it can be argued that authenticity dates back to when we had inauthenticity, which sprung from 19th century adulteration and early 20th century modernization. Such a long period could not be used to describe authenticity as a trend. A trend may grow to become the dominant cultural paradigm, fade away or be incorporated in a compromise hybrid situation. So where are we in this dynamic? Where are we in the hype cycle? Is authenticity a chimaera which when looked at carefully is just a resurgence of an old trope, new wine in old bottles, or a genuine phenomenon of our time. How is contemporary post millennial authenticity different from mid-20th century authenticity and what are the roots and branches of OUR authenticity rather than those other authenticities? To assess this, we need to segment the populations and also the food market. We need to consider generational differences, for baby boomers are not the same as millennials, even though they live at the same time (Johnston and Baumann, 2007).

Demographic transitional factors also play a part as an arbiter of authenticity. Thailand is in a different part of the demographic transition than the EU, and in a different part of the food transformation from locally based agriculture towards city supplied factory production. This transformation leads to increasing obesity, more meat production, more fast food/processed food, and urban food cultures rather than peasant and forest fed agrarians living off their land. The USA, in contrast, is harking back with nostalgia to the peasant agrarian past that is being romanticized and is seeing a resurgence of authentic, locavore foods (Carroll and Wheaton, 2009; Giovannucci et al., 2010; Weaver and Petrini, 2012). So the concept of authenticity will be different due to a different society, value system and social construction.

Retail in a farmers' market is not the same as calling into a 7-Eleven convenience store, though the same demographic group might be served, with the same food needs being met. Authentic growers may not be the same as modernist growers, but the end product can still be an authentic bakery or patisserie.

Globalisation is also a factor affecting authenticity. As products become standardized to meet global needs and desires, the idiosyncrasies of localized branding are swallowed up. For example, the chocolate bar known originally as a Marathon in the UK is now sold as Snickers after a decision to globally align the brands by the owners M&M Mars (Lessware, 2006). Many traditional tea blends once sold in the UK can no longer be bought in the country, only in foreign markets, for example Lipton's (C.R., 2014). Authentic products are also 'localized', so that when attempting to buy British authentic confectionery, the recipes are no longer allowed on the American market due to licensing arrangements' restrictive practices. This happened with Cadbury being blocked by The Hershey Company (Handy, 2015).

Other culturally significant brands have been 'bought out' and then taken off the market and altered, e.g. Peek Freans, Huntley and Palmers biscuits (Sen, 2005). Thus, the removal of nostalgic products from the market creates a desire for the halcyon range of yesteryear in older folks. In this case authentic processed food is being sought, which is slightly different from an individual product that may have been removed. A clear example of this is shown in the fictional film Good bye Lenin! (Becker *et al.,* 2004). In the film about the end of East Germany many Eastern German food brands disappear from the traders' shelves to be replaced with Western products. In reality a hankering for the good old days of the DDR has seen a resurgence for several iconic nostalgia products, with East German consumers purchasing them as an authentic part of their socialist realism culture in the form of 'ostalgie' (Blum, 2000).

4.5 Indicators of authenticity

It is difficult to tease apart the strands of authenticity once they are conflated and interwoven with production processes, terroir and ingredient choices. Concomitant other trends make analysis difficult: Is organic essential for authenticity? Fair trade? Place where sold? Cultural milieu as to who is buying or selling it? Can a hipster buy or sell an authentic cupcake as easily as an Amish farmer? Does marketing and branding make a difference? Examination in context of the marketing mix can indicate (in)authenticity. Being critical of marketing and marketers, who may create hyperreality, or engage in misleading practices and misappropriate authentic attributes in product promotion is no bad thing.

However, marketing per se is not fake and false. We don't see it that way. We believe that true authentic products are also marketed and need marketing, and that there are those companies who are dedicated to delivering true authenticity. Finding particular combinations or forms of (in)authenticity indicators can suggest that a product lacks/has authenticity. These indicators should be taken as suggestions of (in)authenticity rather than as definitive criteria. A matrix can be constructed by scoring a product against the authenticity indicators, thus giving an index of authenticity. To do this a benchmark needs to be created. Then the same kind of product or by analogy similar products can be said to be authentic.

4.5.1 Direct indicators of authenticity

Authenticity of the ingredients

Guidelines and norms that industry accepts as authentic often focus on the authenticity of the ingredients. These can include what we term biological authenticity, such as bacterial cultures in blue cheese and the different protected caves where the culture lives, mould on vines that produces the sweetness, not just production methods like the particular way a pastry is folded or served. This means that genetic testing can be used to verify a correct bacterial strain, though this comes closer to stopping food fraud.

Authenticity of the food processing

A food processing technique that has a verifiable impact on the sensory attributes of the food product. Examples are: sourdough bread (produced in a multi-level process without yeast) vs Chorleywood bread (produced in a single-level process with yeast fermentation); Champagne (Champagne method = fermentation in the bottle) vs Sparkling Wine (tank fermentation); Prosciutto di Parma (conservation by air drying) vs Pressschinken (conservation by preservatives) ... Champagne and Prosciutto di Parma derive their authenticity also from the origin and not only the processing method.

4.5.2 Ecological indicators of authenticity

Location and setting as to where sold

This relates to the place aspect of the marketing mix; authentic pub food cannot be sold in a supermarket. Authentic home-cooked food or farm produce should be available in a farmer's market, craft fair and not a supermarket. If products are sold in these non-authentic places, they lose the connection and thus the authenticity with the producer and place of origin. However, marketing can suggest these aspects, so for example a hypermarket redesigning its meat counter to look like a traditional butcher and staff working in there dressed with butcher's aprons and straw boater hats.

Merchandise around the product

For example, the product is sold with accessories that suggest authenticity. Perhaps cracker biscuits are sold in a hamper with origin-linked cheeses. Or a serving suggestion on the packet implies that something has an authentic use. An example are marshmallows that have a point of sale display with evocative nostalgic pictures of a roasting camp fire and yesteryear campers.

Labelling on the product

Use of allied terms as synonyms of authenticity – such as real, genuine, traditional, locally produced, without being locally or traditionally produced. Such a constructed food is exemplified by the 'non-existent place' of the eponymous Lymeswold cheese, which itself was a combination of different cheese producing traditions (Brown 1982). Also terms that have connotations of authenticity to many people such as Granny's or farmers'/farm house. There are also labels that suggest ethnic origins with supposed authenticity, e.g. the English tea names on Mr. Perkins and Lord Nelson private brands. Most egregious of these are Tesco and Aldi supermarkets' 'fictional farms'; Rosendene Farms, Boswell Farms and Ashfield Farm, that not only are not real functioning farms, but actually conceal that multiple producers from multiple countries have provided the product (BBC News, 2016; Levitt, 2016). These indicate inauthenticity when contrasted with the geographical origin labels, or real people like the Stauds which indicate authenticity.

Legal standards or codes

Stakeholders set these for aspects around the product, not just the product per se. Their existence means that public policy and an institutional environment exits around a product. These are not only the arbiters if something is authentic or not, but also help to create the concept of authenticity (a social construct) and how its semantic field applies OR does not apply in particular application cases. Standards may relate to animal welfare, religious practice, qualification of producer, registration, or some other standard set by a standards body that may not fall under the above attributes, for example Newcastle Brown Ale was brewed in a brewery, which upon relocating changed the geographical boundaries by law as to where it could be produced and yet still be authentic. An absence is illustrative; there is no concept of authentic Viennese water, yet there is a concept of authentic Perrier, Evian has also sought PDO status for its water and in Germany there have been many mineral waters that have gained PDO status. So the idea of authentic mineral waters has been created by the legislative process.

Have we evidence of these standards happening and who has been doing it and how successful are they? Food standards have been set by IFOAM (IFOAM, 2005) across the world. Others have set standards: in viticulture, with cheese councils and the like. Various legal bodies such as the EU or the FDA have also been involved in defining if a product can be described as a bread or soft drink or medicine or other kind of consumable. These have established standards as to what should be in there to meet basic criteria. Most countries also have their own local labels and standards connected to the Codex Alimentarius, which originated in idea from the Codex Alimentarius Austriacus in the Austro-Hungarian Empire (Randell, 1995) and was focused on food safety. There are also some relevant ISO standards and WTO requirements that are universally accepted (Vandecandelaere, 2010).

Authenticity of the terroir, the region

Evidence based on authenticity from a tradition that can be traced and contextualized. So this needs an element of time and also a cultural setting/framing. Examples are Feta Cheese, Champagne, Parmigiano Reggiano, Serrano ham.

Product failures such as Lymeswold cheese could be ascribed to a lack of authenticity (Aslet, 2010). The product was not related to any particular location, whereas Anstey's Double Worcester is related to a place, traditional producer and culture. 'Austrian Gouda' or 'Cornish Brie' are not related to the places of traditional production, but have been successful as they are felt to follow authentic recipes. Could all of these cheeses be called authentic? Cornish and Austrian in their names, as a caveat, suggest not. Brie, Cheddar and Camembert are interesting to look at in this respect. When are they authentic and when not? It appears that some product names have crossed from a sub-culture to many cultures, becoming the 'customary term' (Stern, 1999) and no longer carry the same attributes in terms of authenticity ecology.

4.6 Hipster authenticity, A1, A2 and A3 authenticity

When looking at an underground movement, there is a point when it is picked-up enough to cross the threshold and such that it is no longer the preserve of a sub-culture, but part of a trend in the broader culture. The USP of authenticity stops being unique for one company or one product and can be found in a range of products. For example, cupcakes were definitely a trend; with cupcake shoes, marketing of rail tickets and a lifestyle that goes beyond a single kind of cake amongst others in a bakery. We even have dedicated cupcake stores. At some point there was a crossover when cupcakes were not a product like a muffin or hot cross bun, but gained an added value or wider set of associated features. Authenticity likewise has picked up some connotations and associations over the widening usage of the term re food. Authenticity is not just confined to food, but can be seen as referring to other lifestyle aspects (Banet-Weiser, 2012; Glascock, 2015).

We need deeper analyses when talking about authentic food and authenticity as a trend, which may be culturally prevalent and have a slightly different shade of meaning, than authenticity referring to only one product. For example, in the USA the term hipster is thrown about with unclear definitions, but in many ways it seems to refer to aspirations and values held by a certain demographic profile. Hipsters appears to value authenticity. BUT what they often perceive as authenticity is an ersatz culturally appropriated authenticity (the hyperreal authenticity above) that has been adapted to their more mainstream sensibilities. Perhaps this is the best on offer, as with many cultural things what is available is a Hobson's Choice of that inferior copy or nothing. So for example craft beers, bar-b-cue cookouts as done by 'pitmasters' and other similar appropriations of food traditions. So a contrast with the pitmasters can show how this 'hipster authenticity' may be different from a 'true authenticity' grill experience (Table 4.1).

Table 4.1. Differences in grill cultures between hipster authenticity and true authenticity.

Hipster authenticity – pitmaster (Walsh, 2015)	True authenticity – barbecue man (Wei, 2015)
Myron Mixon sees injecting meat with chemicals and flavoring solutions as 'the latest stage in the evolution of barbecue cookery' (Walsh, 2015)	'injecting meat with chemicals and flavoring solutions isn't 'authentic' barbecue' (Walsh, 2015)
'Texas Barbcue in NYC ... stainless-steel contraptions' they use to cook meat (Walsh, 2015)	'the old-fashioned wood-burning pits in Central Texas' (Walsh, 2015)
'New Yorkers want potato salad, cole slaw, and dessert with their meal' **'the food is not the same without the culture'** tastes worse in NYC (Walsh, 2015)	'authentic traditions like serving the sliced meat on a piece of butcher paper with no utensils' **'the food is not the same without the culture'** – tastes better in Texas (Walsh, 2015)
'The national press would have you believe barbecue is dominated by white hipster males' (Lussenhop, 2015; Walsh, 2015)	'blacks, Latinos, and women are involved in the barbecue biz too' (Walsh, 2015)

Table 4.1. Continued.

Hipster authenticity – pitmaster (Walsh, 2015)	True authenticity – barbecue man (Wei, 2015)
'a new interest in barbecue across the country … competition', and for profit businesses' (Walsh, 2015; Wei, 2015)	'the community barbecue – is largely unnoticed by the general public. Community barbecue isn't about competition; it's about bringing people together… held as private parties to remain out of the jurisdiction of health department bureaucrats'(Walsh, 2015). 'This is an older form than restaurant based, and more authentic to the mass outdoor impromptu cookouts. "Barbecue just got bigger and bigger in the 19th century and was more formalized as a social ritual,' Moss says. 'It was standard during elections and became a way for the community to celebrate the Fourth of July in the South. Hundreds to thousands of people would come to roast cows, pigs, and sheep. There would be a big procession of people coming in. They would retire to a shady grove, chop down trees, and make tables and benches out of them." (Wei, 2015)
Barbecue style is syncretic of several traditional elements and new machinery, techniques.	Barbecue style is embedded in a family tradition or regional style developed from a family (Lussenhop, 2015).
Larger restaurants run as corporations – typically moving toward seated, fine dining and waiter service (Lussenhop, 2015). A strong likelihood of an entrepreneur with grade and hierarchy in the work environment, with fewer opportunities for new workers to eventually own the business.	Mom and pop businesses run to make enough money to support the family rather than a corporation to make profits on a grand scale – typically counter service without waiting staff. (Lussenhop, 2015). More of likelihood of a master-apprentice relationship with greater likelihood of cooperative or other models of ownership than corporate.
Leading workers have a more egalitarian relationship with their customers and may be indistinguishable from them in socio-cultural demographic terms.	Leading workers have a servile, serving relationship to their customers. Customers may have a power gradient, and or socio-cultural differences in relationship with workers (Lussenhop, 2015).
Pitmaster – an entrepreneur who runs a restaurant (Lussenhop, 2015), the term suffering cultural (mis)appropriation in the new businesses. The semantic field excludes the older barbecue man.	Barbecue man – a man in the community that runs a bar b cue as part of a large cookout. The term gradually replaced by pitmaster, but in a way so that the semantic fields overlap with barbecue man, possibly being used interchangeably.

Nevertheless, despite these differences we could still describe both as being linked to an authenticity trend. We might give a level of authenticity, perhaps A1, A2 or A3 (these could be applied for each indicator, but the idea is that they give an aggregate level, much as the European Common Framework of languages says someone has B1 or B2 overall level of ability in a language, rather than B1 in reading comprehension and B2 in spoken production and C1 in listening comprehension). Just as speakers of a language are described as L1 (native speakers), L2 (proficient second language users) and L3 (restricted and partial users of a language) we can say in our case A1 would be embedded in the cultural milieu and would be the traditional practices the EU supposedly tries to protect with its PDO-PGI-TSG scheme. A2 would be the altered forms as found in manufacturing adaptations of those traditions. Craft authenticity (Carroll and Wheaton, 2009) would fall into A2 as it is recognizable, but may display evidence of cultural appropriation as it has been adapted by those from another background than in A1 authenticity. A3 would contain elements of such authenticity, but in no way would they be considered as genuine-bona fide originals.

4.7 Authenticity in the food chain

So far we have talked about a product being 'authentic' from the consumer marketing perspective, i.e. as a finished product. If we look at the whole human food chain for a product we find there are suppliers, growers and manufacturers with many end products made from several ingredients. We can look at an individual ingredient and talk about authenticity here. The EU Food Integrity Project is looking here at authenticity in a different way (Spink and Moyer, 2016). Authenticity in the food chain is seen as avoiding food fraud and is better termed food integrity (Hoorfar *et al.*, 2011). We authenticate something as a verb, it has been authenticated; rather than an adjective, as an authentic thing. 'Food fraud is a collective term used to encompass the deliberate and intentional substitution, addition, tampering, or misrepresentation of food, food ingredients, or food packaging; or false or misleading statements made about a product, for economic gain' (Spink and Moyer, 2011).

A case in point would be the meat dish Shepherd's Pie sold as containing lamb, when in fact horse meat had been used. While claiming it is lamb, when in fact it was horse would be obvious to anyone as fraudulent. So the question rises, if a horse meat shepherd's pie is inauthentic just because it doesn't contain lamb. Historically shepherd's pie could contain lamb or beef and be named 'Shepherd's Pie' (Ayto, 1990). A meat-identifying label was not needed in the name. In more recent times claims have been made that it should only contain lamb, this is a folk etymology, and beef has long been acceptable, though the dish is often called Cottage Pie if containing beef. When horse meat was used it has been called Stable Hand's Pie (Windle, 2013). And yet Lentil and Wild Mushroom Shepherd's Pie and Turkey

Sweet Potato Shepherd's Pie also exist (Gordon, 2012). So here the inauthenticity was not the use of horse per se, but the incorrect labelling.

In the UK the Trading Standards is responsible for monitoring local food establishments and checking that what is marketed is what is sold. A common finding in pizzerias is that a mozzarella cheese pizza in fact is not 100% mozzarella cheese. Instead it has been adulterated with a share of rennet. You should be justly suspicious of 'cheese' topped pizzas where the cheese topping is an agglutinated mass and solidified rather than a softer, stretchier cheese. Another example is found in the marketing of *petits pois*. *Petits pois* translated from French means small peas. So many people think that *petits pois* are ANY small peas. But in fact *petit pois* is a particular variety of pea. The average consumer buying *petits pois* would not be able to identify this discrepancy. Another is Wasabi. Wasabi is a root vegetable and to maintain its flavour was grated with a sharkskin grater called a *samegawa* or a copper grater called *oroshigani*. When this happens you can be fairly sure that you are getting real wasabi. However often wasabi as sold outside of Japan is not wasabi, but instead horseradish with a green colorant and a token amount of wasabi oil, plant or other substance so that wasabi can be written on the packaging.

In these cases, authenticity is missing at the ingredients level, but the marketing and labelling and appearance of the food may appear authentic to a consumer. The demand for buying authentic products can be satiated with these methods. In fact, perversely, there are cases of food colorants, flavourings or flavour enhancements being seen as more authentic than true examples of a product. These findings must be treated with caution as individual memories are faulty, people may not have had as much exposure to authentic products as they think they have and preferences can have changed due to long exposure to inauthentic food products (Kuhn, 2013). For example, blander foods that preserve longer will be less flavoursome than freshly prepared foods. A good example of this is wheat products. Freshly ground wheat flour used for baking and eaten the same day tastes noticeably different from the same products made from flour that has been kept for longer, even if consumed shortly after baking. The freshly baked bread of today is not the same as the freshly baked bread of our Victorian forbearers (Barber, 2014).

Novel foods and foods made using genetic engineering are not authentic on many grounds, but when argued as substantially equivalent may be formulated as an ingredient in an authentic end product. Here traceability and transparency are required to meet the non-fraudulent authenticity criteria. Consumer perceptions that such foods are misrepresented and not authentic are based on the novelty aspect being given a greater weighting than other aspects, which leads them to conclude that there is food fraud (Spink and Moyer, 2011).

4.8 Fakelore and imagined authenticity

Authenticity as a trend does not mean that a product is truly, genuinely really an authentic product in every case. Authentic products can also be 'authentic' or true to an image or a suggested reality that never existed. Creating that bogus reality is done routinely by supermarkets. For example, Austrian discount retailer Hofer markets their meat products as if they are prepared by a small family butcher. This 'family butcher archetype' can be seen in adverts made to be shared on social media like YouTube. This is not necessarily a hyperreality, but one where otherness is fetishized and created (Deepdas, 2013). Other products suggest an authenticity that is ersatz and if looked at from reality are meaningless. In the case of Lymeswold this consisted of the Milk Marketing Board using the fictitious village of Lymeswold as a setting for the cheese's back story (Aslet, 2010; Brown, 1982), with associations desired with the 'heart of England' (Brown 1982) 'bucolic traditionalism' (Ayto, 2004). 'Farmer's' or 'Kingsmill' or 'Village' are used to suggest feelings and connotations by association with the wider semantic feel of those key words, rather than giving a true authentic heritage. This practice is called folklorismus (Newall, 1987) or fakelore (Smith, 2001).

Folklorismus or folklorism can have slightly different meanings, when historically 'Scholars west of the Iron Curtain usually identified folklorism in a commercial context, while those to the east identified folklorism in government sponsored cultural programs' and both may be labels applied to material culture or to processes (Šmidchens, 1999). In the case of processes, it has been argued that 'authenticity [of the product intrinsic characteristics] is irrelevant to the ethnographic description of these processes' (Šmidchens, 1999). In application to food, (food items also being a part of material culture (De Solier, 2013)), Smith (2001) proposes a typology:

- Journalistic enrichment: … Journalists … add facts, dialogue and drama…they frequently rely on interviews and often fail to check … with primary sources… Logical fallacies and presentism [are often found].
- Culinary jingoism: undocumented stories about cuisines intended to improperly attribute origins of particular dishes to specific location, group or nationality… Local boosterism… often to attract tourists…Negative stereotypes…Temporal jingoism: the belief that food and cookery in the present are better than food and cookery in previous times.' The latter being a manifestation of the enlightenment idea of progress as chronological snobbery, which is a form of the logical fallacy of appeal to novelty.
- Great (usually white) men stories: undocumented attribution of origins of dishes to specific individuals…undocumented personal claims by individuals to have invented dishes when they did not do so… famous personage dishes…historical revisionism… invented culinary traditions [like] 'The Soul Food' tradition of African-Americans.

- ▶ Commercial promotion: undocumented stories repeated by corporations in their advertising intended to sell commercial products. Almost all food advertising that makes reference to historical events is inaccurate. In this case, advertisers are interested in promoting sales not in accurately telling history.
- ▶ Health myths: medicinal claims for specific foods/products frequently repeated without solid scientific basis. Vegetarian/anti-vegetarian myths… Aphrodisiac claims.' These seem to be related to folk-etymology and old wives' tales such as tomatoes called love-apples mean that people in the past regarded them as aphrodisiacs, where there is scant 'evidence for this claim except in modern literature.

4.9 Conclusions

When we talk of authenticity it perhaps has a much wider field of application than some other trends. For example, fair trade and buy local have global implications, but are clearly delimited. To manageably research authenticity as a trend we can look at several food sector segments such as barbecue food, preserved food and bakeries in concert, rather than any one part of the food cultural landscape alone. The concept of cultural omnivorousness where there is 'cultural openness to multiple ethnic and class cuisines' (Johnston and Baumann, 2007) when applied to food makes this requisite if we look to a trend rather than specific food examples.

Geographical limitations are also needed to identify where authenticity is talked about, which are complicated by the effects of globalization, making everything available anywhere. 'Authentic' in referring to food has been happening since the 1940s (Johnston and Baumann, 2007), so requires time-wise delimiting in the context of any research. However, by using clear metrics, reliable research from sales data and our taxonomies it is possible to identify an authenticity trend in time and space in a given market.

Yet, at the moment the literature is unclear and conflates: traditional foods; hipster or artisanal production; hyperrealism, rural or urban; class, particularly working class; historically accurate or nostalgic foods; and ethnic foods under the label of authentic. The cultural milieu that makes a food authentic, the processes of marketing, production methods and possibly even the ingredients are further aspects in the mix. Ironically, it is possible to have an authentic food that lacks a defining ingredient, or lacks some other authentic ecological element, an authentic mock turtle in our mock turtle soup! Similarly, there are many terms that are used for authentic synonyms so that a clear definition cannot easily be given without exploring the semantic fields and cultural application of those terms in context.

References

Adria, F., Blumethal, H., Keller, T. and McGee, H., 2006. Statement on the 'new cookery'. The observer. Available at: http://tinyurl.com/yavgxr74.

Alsip, M.A., 2015. The 'Food Babe': a taste of her own medicine. Skeptical Inquirer 39: 39-41.

Aslet, C., 2010. Blessed are the British cheesemakers: booming exports prove how much we've matured since the days of the awful Lymeswold, says Clive Aslet. The Telegraph, 23-9-2010. Available at: http://tinyurl.com/yath8h95.

Ayto, J., 1990. The Glutton's glossary: a dictionary of food and drink terms. Routledge, London, UK, 323 pp.

Ayto, J., 2004. An AZ of food and drink. Oxford University Press, Oxford, UK, 388 pp.

Banet-Weiser, S., 2012. AuthenticTM: the politics of ambivalence in a brand culture. NYU Press, New York, NY, USA, 279 pp.

Barber, D., 2014. The third plate: field notes on the future of food. Penguin, New York, NY, USA, 496 pp.

BBC News, 2016. Tesco 'fictional farms' may be misleading. Available at: http://www.bbc.com/news/business-35889282.

Becker, W., Brühl, D., Sass, K., Khamatova, C., Arndt, S., Lichtenberg, B. and Pool, X.F.C., 2004. Good bye Lenin! Cameo Media, London, UK.

Bell, D. and Hollows, J., 2011. From river cottage to chicken run: Hugh Fearnley-Whittingstall and the class politics of ethical consumption. Celebrity Studies 2: 178-191.

Bell, M.M., Lloyd, S.E. and Vatovec, C., 2010. Activating the countryside: rural power, the power of the rural and the making of rural politics. Sociologia Ruralis 50: 205-224.

Benjiw, G., 2011. Cho Cho Apple Pie. Jamaican Foodie. Available at: http://tinyurl.com/s58edm6.

Blum, M., 2000. Remaking the East German past: ostalgie, identity, and material culture. Journal of Popular Culture 34: 229-253.

Bogle, E., 1993. Plastic paddy. Mirrors. Larrikin Music Pty Limited, Sydney, NSW, Australia.

Brown, E., 1982. First new cheese in Britain in 300 years. The New York Times, 27-10-1982. Available at: http://tinyurl.com/sy3cqag.

C.R., 2014. Why Guinness is less Irish than you think. The Economist explains: explaining the world, daily. Available at: http://tinyurl.com/kgbgkla.

Carroll, G.R. and Wheaton, D.R., 2009. The organizational construction of authenticity: an examination of contemporary food and dining in the US. Research in Organizational Behavior 29: 255-282.

De Solier, I., 2013. Food and the self: consumption, production and material culture. Bloomsbury Publishing, London, UK, pp. 224.

Deepdas, S., 2013. The other in the space of an-other. Helsinki University, Helsinki, Finland.

DeMers, J. and Fuss, E., 1998. The food of Jamaica: authentic recipes from the jewel of the Caribbean. Simon and Schuster, New York, NY, USA, pp. 132.

Dougherty, D., 2012. The maker movement. Innovations 7: 11-14.

Douglas, M., 2014. Food in the social order. Routledge,Oxford, UK, pp. 304.

Edmundson, J. and Martusewicz, R.A., 2013. Putting our lives in order. In: Kulnieks, A., Longboat, D.R. and Young, K. (eds.) Contemporary studies in environmental and indigenous pedagogies: a curricula of stories and place. SensePublishers, Rotterdam, the Netherlands, pp. 322.

Ellis, C., 2008. Meeting King Corn: Earl Butz was a product of his time. Culinate. Available at: http://archive.is/BpQOb.

Forlizzi, J., 2008. The product ecology: understanding social product use and supporting design culture. International Journal of Design 2(1).

Gilmore, J.H. and Pine, B.J., 2007. Authenticity: what consumers really want. Harvard Business Review Press, NewYork, NY, USA, pp. 320.

Giovannucci, D., Barham, E. and Pirog, R., 2010. Defining and marketing 'Local' foods: geographical indications for US products. Journal of World Intellectual Property 13: 94-120.

Glasock, T., 2015. Hipster barbie is so much better at instagram than you. Available at: http://tinyurl.com/vztb57r.

Gordon, M., 2012. A new spin on an old classic: 5 recipes for Shepherd's Pie. Apartment Therapy, Seattle, USA. Available at: http://tinyurl.com/y9h7qvy4.

Grunert, K.G. and Aachmann, K., 2016. Consumer reactions to the use of EU quality labels on food products: a review of the literature. Food Control 59: 178-187.

Gura, T., 2001. The battlefields of Britain. Nature 412: 760-763.

Ha, T., 2008. The Australian green consumer guide: choosing products for a healthier home, planet and bank balance. University of New South Wales Press, Sydney, Australia, pp. 239.

Handy, B., 2015. The bitter, not-sweet cadbury-chocolate war. Vanity Fair, May 2015 edition. Available at: http://tinyurl.com/s55thcw.

Hoorfar, J., Jordan, K., Butler, F. and Prugger, R., 2011. Food chain integrity: a holistic approach to food traceability, safety, quality and authenticity. Elsevier, Amsterdam, the Netherlands, pp. 384.

International Federation of Organic Agriculture Movements (IFOAM), 2005. Principles of organic agriculture. IFOAM, Bonn, Germany. Available at: http://tinyurl.com/kurzckj.

Johnston, J. and Baumann, S., 2007. Democracy versus distinction: a study of omnivorousness in gourmet food writing. American Journal of Sociology 113: 165-204.

Kuhn, E.M.B., 2013. Tirolerknödel in Brasilien-Weitergabe und Veränderung von kulinarischem Wissen bei Tiroler Migranten und Migrantinnen in Brasilien. Dissertation, University of Vienna. Available at: http://tinyurl.com/swwxjc5.

Lessware, J., 2006. Is Heinz's famous product in danger of becoming another has-bean? The Scotsman. Available at: http://tinyurl.com/qp8ke9q.

Levitt, T., 2016. Tesco's fictional farms: a marketing strategy past its sell-by date? Available at: http://tinyurl.com/z4xph49.

Levy, H., 2015. What's new in Gartner's hype cycle for emerging technologies. Gartner Inc. Available at: http://tinyurl.com/petb6w4.

Lussenhop, J., 2015. Black pitmasters left out of US barbecue boom. BBC News Magazine Online. Available at: http://tinyurl.com/ya8okm83.

Mandal, P.K.E.I.C., 2016a. Portal journals agriculture and food sciences JFNR home aims and scope. Available at: http://www.sciepub.com/journal/JFNR/aimsandscope.

Mandal, P.K.E.I.C., 2016b. Portal journals agriculture and food sciences JFNR. Home – Frequently Asked Questions. Available at: http://www.sciepub.com/journal/JFNR/faq.

Mannur, A., 2009. Culinary fictions: food in South Asian diasporic culture. Temple University Press, Philadelphia, PA, USA, pp. 255.

Munoz, C.L., Wood, N.T. and Solomon, M.R., 2006. Real or blarney? A cross-cultural investigation of the perceived authenticity of Irish pubs. Journal of Consumer Behaviour 5: 222.

Newall, V.J., 1987. The adaptation of folklore and tradition (Folklorismus). Folklore 98: 131-151.

Oliver, T., 2013. The real coke, the real story. Random House Publishing Group, New York, NY, USA, pp. 208.

Pendergrast, M., 2013. For God, country, and coca-cola. Basic Books, New York, NY, USA, pp. 560.

Pirker, H., Haselmair, R., Kuhn, E., Sschunko, C. and Vogl, C.R., 2012. Transformation of traditional knowledge of medicinal plants: the case of Tyroleans (Austria) who migrated to Australia, Brazil and Peru. Journal of Ethnobiology and Ethnomedicine 8: 44.

Randell, A., 1995. Codex Alimentarius: how it all began. FAO, Rome, Italy. Available at: http://tinyurl.com/rauhp9a.

Scully, M.D., 2009. Plastic and proud?: discourses of authenticity among the second-generation Irish in England. Available at: http://tinyurl.com/wmxsa4w.

Sekules, V., 1980. Friends of the earth cookbook. Penguin Books, Middlesex, UK, pp. 192.

Sen, S., 2005. Indian spices across the black waters. In: Avakian, A.V. and Haber, B. (eds.) From Betty Crocker to feminist food studies: critical perspectives on women and food. University of Massachusetts Press, Amherst, MA, USA, pp. 185-99.

Šmidchens, G., 1999. Folklorism revisited. Journal of Folklore Research 36: 51-70.

Smith, A.F., 2001. False memories: the invention of culinary fakelore and food fallacies. Food and Memory Proceedings of the Oxford Symposium on Food and Cookery 2000, 2001. Oxford, UK, pp. 254-260.

Smith, K.K., Gabriele, K., Kornfeld, S., Märzendorfer, E., Rösch, C. and Foszczynski, M., 2014. Vor Ort zu Besuch bei fünf produzierenden Unternehmen in Wien. ON Das Magazin der Wiener Wirtschaft Vienna, Austria: WKO Wirtschaftskammer Wien. Available at: http://lesen.wkw.at/on/pdfs/1290.pdf.

Spink, J. and Moyer, D.C., 2011. Defining the public health threat of food fraud. Journal of Food Science 76: R157-R163.

Spink, J. and Moyer, D.C., 2016. Introducing the Food Fraud Initial Screening model (FFIS). Food Control 69: 306-314.

Stern, A., 1999. The protection of geographical indications in South Africa. Symposium on the international protection of geographical indications. World Intellectual Property Organization, Somerset West, Cape Province, South Africa.

Vandecandelaere, E., Arfini, F., Belletti, G. and Marescotti, A., 2010. Linking people, places and products, a guide for promoting quality linked to geographical origin and sustainable geographical indications. FAO, Rome, Italy. Available at: http://tinyurl.com/reurz57.

Walsh, R., 2015. 7 dirty truths about BBQ (that nobody wants to talk about). First we feast. Available at: http://tinyurl.com/tdof8ep.

Weaver, J. and Petrini, C., 2012. Locavore adventures: one chef's slow food journey. Rutgers University Press, New Brunswick, NJ, USA, pp. 240.

Wei, C., 2015. An illustrated history of barbecue in America: author and southern living BBQ editor Robert Moss explains how the ancient tradition of smoking meat over charcoal grew into a distinct form here in the USA. First we feast. Available at: http://tinyurl.com/v5z35dl.

Whitley, A., 2011. Bread matters: the state of modern bread and a definitive guide to baking your own. Andrews McMeel Publishing, Kansas City, MO, USA, pp. 416.

Windle, C., 2013. The horsemeat cookbook. Random House Publishing Group, New York, NY, USA, pp. 128.

Zlotnick, S., 1996. Domesticating imperialism: curry and cookbooks in Victorian England. Frontiers: a Journal of Women Studies 16: 51-68.

Consumer trends and new product opportunities in the food sector

5. The convenience and bundling trends

Ľ. Nagyová and *I. Košičiarová*

Slovak University of Agriculture in Nitra, Tr. A. Hlinku 2, 949 76 Nitra, Slovakia; ludmilanagyova@hotmail.com

Abstract

The globalized market environment of today is characterized by turbulent development of science and technology stronger and sharper linking between different market environments, economies and populations, increasing pressure on people and their performance higher demands for a healthy lifestyle and protection of the environment; and shortening of the time spent with family and friends, which can be seen also in the fact that the consumers are seeking for solutions which allow them to maximize their free time and spend more time doing the things which they like and value. This is why the convenience trend is driven by this desire to create more leisure time. Convenience is multi-faceted and will continue to evolve and converge with other trends to meet the customer's needs and the demands of society. The convenience trend is supplemented by the bundling trend. 'Bundling plays an increasingly important role in many industries and some companies even build their business strategies on bundling.' (Fuerderer *et al.*, 2013) Bundling is based on the convenience trend, because the customers can save time by buying two or more products packed together instead of having to think and buy them separately. To rise to the challenge of meeting customers' need for greater convenience, retailers must offer products that provide speed and ease.

Keywords: convenience, convenience food, bundling, trend

5.1 Convenience, its definition, basic forms, factors leading to it, and benefits

Convenience is an attribute of a food product for which the demand is increasing (Buckley *et al.*, 2005). The result of many studies done on the issue of convenience is that convenience is just as important as attributes like taste, health and price in determining a consumer's preference towards a food-product (Candel, 2001). We can find many different meanings of the word convenience, referring to time utilisation, accessibility, portability and handiness. The increasing importance of convenience in food products is manifested in the development of selected product lines or ranges (Frewer *et al.*, 2013):

- ▶ The greater importance of both frozen and chilled products and dried ready-made products is noteworthy – the brands from companies such as Unilever, Dr. Oetker or Nestlé are the market leaders within these product groups in Europe.
- ▶ Predominantly fresh ranges, which are generally not branded but are strongly influenced by the retailer, e.g. salads, sandwiches or other prepared products – these fresh convenience ranges correspond to the desire of retailers to add value, which shows that both retailers and manufacturers of branded goods have recognised the increasing significance of convenience and endeavour to position their brands in this sector.

> Today time has become one of our most precious assets, and many consumers therefore buy products at the closest location to preserve time for other activities. Even a buyer who prefers a specific brand will readily choose a substitute if the preferred brand is not conveniently available.
>
> Pride and Ferrell, 2008

Convenience in the context of food can be defined in terms of reductions in time and effort (even if it is mental or physical effort) spent by buying, storing, preparing and consuming the food (Recapt, 2011) Convenience foods are then defined as any fully or partially prepared food in which significant preparation time, culinary skills or energy inputs have been transferred from the homemaker's kitchen to the food processor and distributor Buckley *et al.* (2005).

Nowadays, many people feel constrained to eat so called convenience food as a response to intransigent problems of scheduling everyday life (Warde, 1999):

> A distinction is drawn between modern and hypermodern forms of convenience, the first directed towards laboursaving or time compression, the second to time-shifting. It is maintained that convenience food is as much a hypermodern response to de-reutilization as it is a modern search for the

reduction of toil. Convenience food is required because people are too often in the wrong place; the impulse to time-shifting arises from the compulsion to plan ever more complex time-space paths in everyday life. The problem of timing supersedes the problem of shortage of time.

(Srinivasan and Shende, 2014)

Convenience foods must be tasty and high quality, while meeting consumer expectations in terms of ease of use, safety, variety, packaging, nutritional value and product appeal. And consumers are attracted to them by their relatively low cost and complexity, ease of use, variety, availability despite the season, and food safety.

(Cargill, 2017)

Convenience food can include products such as candy, beverages such as soft drinks, juices and milk, fast food, nuts, fruits and vegetables in fresh or preserved states, processed meats and cheeses, and canned products such as soups and pasta dishes (Figure 5.1). Additional convenience foods include frozen pizza and cookies (Rees, 2005), or chips and pretzels (Chow, 2007). These products are often sold in portion-controlled, single-serving packaging designed for portability (Frewer *et al.*, 2001; Rudolph *et al.*, 2012) and they can be sold as hot ready-to-eat meals, room temperature products or as refrigerated or frozen products that require little preparation (usually microwave or oven-ready).

The convenience trend symbolizes the increased time pressure, stress and work-life balance problems that consumers are experiencing, which are also shown in Figure 5.2.

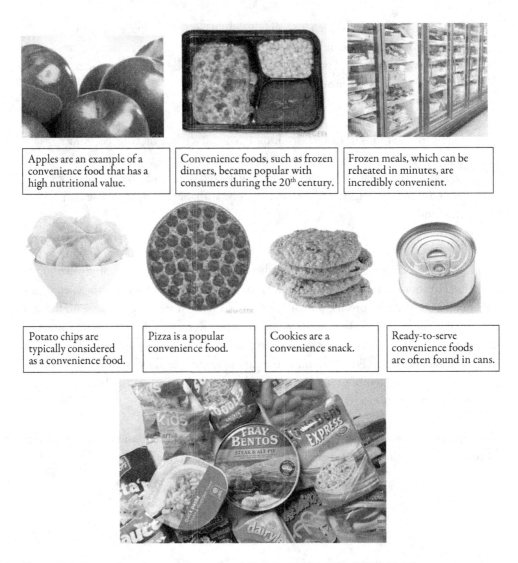

Apples are an example of a convenience food that has a high nutritional value.

Convenience foods, such as frozen dinners, became popular with consumers during the 20th century.

Frozen meals, which can be reheated in minutes, are incredibly convenient.

Potato chips are typically considered as a convenience food.

Pizza is a popular convenience food.

Cookies are a convenience snack.

Ready-to-serve convenience foods are often found in cans.

Figure 5.1. Examples of convenience food (Boseley 2010; WiseGEEK, 2016).

All of the above-mentioned factors, but also the fact that present life is much faster than it was a few years ago, result in that people have less time that they could spend with their families and friends, on leisure activities, but also on cooking. This is why customers seek for fast and convenient solutions, which can save time. As examples of 'time saving' options could be mentioned the following (Consumer Trend Report, 2010):

▶ Simplified meal preparation: there is an ongoing market for products and services, which cater to those who do not have the time or skills to prepare complex meals, but they still

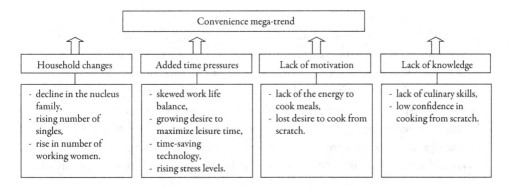

Figure 5.2. Factors leading to the convenience (Consumer Trend Report, 2010).

want nutritious options. This is why there exists numerous cookbooks and TV programs based on the theme of simple and healthy meal solutions (e.g. Jamie Olivier's 15 minutes meals).

► Prepared meal: consumers can buy prepared meals that only require heating in the microwave or oven, or quick assembly, such as bagged salads. These meals offer the consumers a wide range of foods, which they can enjoy without the difficulty of following long or complex recipes. There exists numerous options for fully prepared meals including ready-meals, pizzas and take-away food for consumption at home; or for so called partly prepared meals, which consist of several meal components, such as wet and dry sauces.

► Food on-the-go/ ready-to-eat meals: it refers to a meal/food, which can be consumed in the same state as in which it was sold or distributed (Bae *et al.*, 2010) and it can be characterized as the food which is suitable to be consumed while driving to and from the work, while working, or while engaging in leisure activities.

► Online grocery shopping: new trend which is nowadays still less used as classical forms of purchase, but which is fighting for its place in the market. Consumers are seeking for faster ways to purchase, which can be done from their homes.

Some companies have got beyond these basic time-saving options and they have come up with novel approaches, which include (Consumer Trend Report, 2010):

► Meal assembly concept: this concept appeared for the first time in the USA in the early 2000s as a response to the convenience and health requirements of consumers. Stores allow the customers to procure a month's supply of evening meals at a local franchised outlet in a one or two-hour session, using pre-cut and pre-measured ingredients and following predetermined recipes. This concept satisfies two key consumer demands of suburban working moms – health and convenience.

- ▸ Delivery of organic products: with this system farmers, small businesses but also some supermarkets capitalize on consumer demand for organic and ethical food, which is locally sourced and delivered directly to the required place (home, workplace, etc.).
- ▸ Customized foods for children: as an example can be mentioned the food retail and eating outlet 'Kidfresh', which can be found in New York (Figure 5.3). This retail outlet is designed especially for children and it caters to affluent and busy families who desire healthy convenience food. The outlet concentrates on healthy foods and markets products as being largely organic, free of artificial preservatives and additives, and low in fat and sugar. The concept capitalizes on both the health and convenience trends.

Box 5.1. Futuristic cooking methods (Consumer Trend Report, 2010).

In the future, convenience could mean consumers growing their own food. The Cocoon cooker was the winner at a recent Electrolux Design Contest. The Swedish invention by Rickard Hederstierna is an example of forward-thinking technology. It is a food generator that lets the consumer grow his or her own meat. According to Hederstierna, the Cocoon could make food shortages a thing of the past. 'This will create 100% pure meat without the need for animals to be killed and with no risk of contamination,' said the 27-year-old. 'It will change everything.' The design, described as 'controversial' by one of the judges, is made of glass, and works by cooking pre-mixed packets of muscle cells, oxygen and nutrients.

Figure 5.3. Kidfresh signature sandwich (the Shapewich) and the storefront (Consumer Trend Report, 2010).

Table 5.1. Benefits and drawbacks of convenience food (based Bodyecology, 2015; Cardiff, (2013; Srinivasan and Shende, 2014).

Benefits of convenience food	Drawbacks of convenience food
Convenience food reduces the time required for preparation and/or the cooking time.	Some convenience food provide only little of nutritional value.
Due to the various techniques used in the convenience food manufacturing process, one is able to get a lot of variety in the market. In fact, the amount of variety available in the stores today does put a working woman in a fix on the choice that she needs to make. Another point, which has to be noticed, is that this variety remains constant in the market and does go off shelf citing seasonal reasons like the fresh fruits and vegetables do.	Most of convenience food have excessive amounts of sodium, sugar, saturated fats.
One can read the contents and detailing on the packaging of the convenience product and decide to whether health aspects are preferred or not.	Processed food is loaded with preservatives (e.g. MSG), unnatural colouring, added flavouring, etc.
Convenience products are convenient to carry, stack and store since their packaging is well planned considering various factors like consistency, texture, size and fragility.	Some processed food is filled with indiscernible parts and pieces.
Most convenience foods have a longer shelf life due to additives and the same is mentioned on the package hence the user is well informed of its usage period.	To make up for the loss of nutrients during processing, synthetic vitamins and minerals are added to boost their nutritional content.
Storage of convenience foods is easy as they can be stacked up anywhere and can also help better utilization of the space available. Regeneration is also faster and better.	Regular eating of processed food can increase the risk of diseases (e.g. cancer, obesity, diabetes, etc.).
Convenience food helps saving labour in terms of going to market for purchase, pre-cleaning, preparation and post preparation cleaning.	
The new techniques used in the manufacturing and packaging of convenience like aseptic canning, rapid freezing, various methods of eradicating bacteria reduce spoilage of food if stored properly.	

Table 5.1. Continued.

Benefits of convenience food	Drawbacks of convenience food
Many a time the working woman does have requests from her own home or she herself wants to try preparing something of which the recipe might not be known for her. Hence, she looks out for such products for which she need not bother to know how to prepare it rather she would use the convenience of such available products.	
Certain fresh products might not be available throughout the year due to their seasonal availability whereas convenience products are manufactured using stringent quality standards hence their availability besides being consistent also has consistent taste, texture and taste.	

The basic possibilities how to lower or avoid the salt content of convenience foods are for example the purchase of products which say 'with no salt added'; usage of fresh poultry, fish and lean meat, rather than canned or processed types; cooking of staples such as rice, pasta and hot cereal without salt; choice of those convenience food which are low in sodium; avoiding of frozen dinners, pizza, packaged mixes, canned soups and salad dressings, because they contain often a lot of sodium; and rinsing the canned food, such as tuna or beans, to remove some sodium (NutritionVista, 2009).

5.2 Bundling, its definition, basic forms and benefits

The most common definition of bundling comes from the year 2002 and it says that bundling is a sale of two or more separate products in one package. Its formulation however was not simple.

The first work on bundling dates back to the 70s of the 20th century, when Adams and Yellen defined bundling as the act of 'selling goods in packages'(Adams and Yellen, 1976). While this definition is very simple and easy to be remembered, there is a degree of haziness around the attributes of 'goods' and 'packages'. In the 1980s, Guiltinan introduced a further attribute, namely the 'price'. In his view, bundling can be defined as 'the act of marketing two or more products and/or services in a single package for a special price' (Guiltinan, 1987). The underlying idea is that a reduction is given when a customer buys several products at the same time (Chiambaretto and Dumez, 2012).

Further, Yadav and Monroe (while they want to establish a brief framework in the introduction before going on to a mathematical proof of the superiority of bundling) emphasize the attribute of 'tying' and they define bundling as 'the selling of two or more products and/or services at a single price' (Yadav and Monroe, 1993).

The final and most relevant definition of bundling is the one given by Stremersch and Tellis, who say that bundling is 'the sale of two or more separate products in one package' (Stremersch and Tellis, 2002). Here, the concept of price disappears, but another much more important attribute is emerging – 'separate products'. This definition places the emphasis on the strategy in its dynamic dimension – it involves combining two or more existing products which have been until now marketed separately (Chiambaretto and Dumez, 2012).

We must also mention that there is a difference between price bundling and product bundling (Table 5.2). While price bundling is the sale of two or more separate products in a package at a discount, without any integration of the products, product bundling can be defined as the integration and sale of two or more separate products in one package at any price (Stremersch and Tellis, 2002). It is also about the creation of differentiation, greater value and therefore enhancing the offering to the customer. This approach is radically different from the price one, since it is assumed that the integration of products will create value for customers, for example by reducing the risk of incompatibility (Telser, 1979). Bundling is based on the idea that consumers value the grouped package more than the individual items (Berry, 2012).

A bundling strategy can be used not just for product repositioning, which means the altering of an existing product to make it more appealing to the market place, but also for product differentiation, in which one or more attributes of an existing product are modified to make it different from others. This means that the bundling of an existing product with an additional feature/features or services can serve as a repositioning or differentiation strategy (Breidert, 2005). As an example can be mentioned the Coca-Cola company, which always tries to come to the market with something new, or attractive to the customer. In Figure 5.4 and Figure 5.5 we show two different examples of Coca-Cola's bundling strategy. While Figure 5.4 shows a design that Coca-Cola used during the Fifa World Cup (thus bundling Coca-Cola with the FIFA World Cup, where we can see the name 'New Mini Football Bottles' and atrophy besides the Coca-Cola signature, as well as the changed layout of the Coca-Cola – circle shape to act like a football), Figure 5.5 shows an advertisement bundling Coca-Cola with Japanese seasoning (it is a seasoning advertisement bundle, where the package is red colour, which is representing Coca-Cola and many flowers on the tin, which mean a spring festival limitation design) (Kah wai, 2014).

Table 5.2. Bundling terms (based on Stremersch and Tellis, 2002).

Term	Definition	Examples
Bundling	Bundling is the sale of two or more separate products in one package.	Multipacks of beer, juices, soft drinks; 1 kg pack of apples
Price bundling	Price bundling is the sale of two or more separate products as a discount, without any integration of the products.	Variety pack of cereals, lunchtime menu in a restaurant
Product bundling	Product bundling is the integration and sale of two or more separate products at any price.	Starbucks
Pure bundling	Pure bundling is a strategy in which a firm sells only the bundle and not (all) the products separately.	
Mixed bundling	Mixed bundling is a strategy in which a firm sells both the bundle and (all) the products separately.	Fast food or cinema combos

Figure 5.4. Example 1 of bundling in the case of Coca-Cola (Kah wai, 2014).

Figure 5.5. Example 2 of bundling in the case of Coca-Cola (Kah wai, 2014).

Even if the issue of bundling is not so new, until now there have been only few studies to determine whether consumers actually prefer bundles, if the sales increase is a direct result of price bundling, and if the bundling cannibalizes the sales of standalone products. A study by Derdenger and Kumar on the video game market concluded that the bundles can entice consumers to buy consoles earlier, especially when they are offered the choice of buying the bundle or just the console, which means that mixed bundling, is better than pure bundling (Derdenger and Kumar, 2013).

As it is shown in Figure 5.6, we can distinguish between several types/forms of bundling (Fuerderer *et al.*, 2013):

► Separate pricing / pure component pricing: products are offered and priced individually.
► Pure bundling: only the bundle is sold, which means that the products cannot be bought individually.
► Mixed bundling: both the bundle and the individual product are offered and sold, i.e. a combination of separate pricing and pure bundling. This form of bundling is well used by McDonald's, which uses it for its value meals. In this form of bundling a discount can also be given for the second product if the full price is paid for the first/leader product. This form is then called 'mixed-leader bundling', where the 'leader product' is priced high and innovative, while the second product is priced low and mature.
► Tie-in bundling: a special form of bundling, where the buyer of the main product (tying good) agrees to buy one or several complementary goods (tied goods), which are necessary to use the tying good, exclusively from the same supplier.
► Add-on bundling: this form of bundling is not strictly a mixed bundling option, because an add-on product will not be sold unless the leader product is purchased. This is why

the add-on bundling is similar to tie-in sales. As an example of add-on bundling can be mentioned a lunchtime menu with an extra dessert.
► Sales rebates: companies are frequently offering to their customers so called year-end rebate on the total annual sales across all the company's products. Sales bonuses are a mixture between bundling and nonlinear pricing, because it does not matter whether the total sales come from one or from several products.
► Cross couponing: coupons are frequently used for the promotion of products from consumer-goods manufacturers company's assortment. E.g. Coca-Cola reached diet soft-drink users by the distribution of coupons for Diet Minute Maid on its 2-liter Diet Coke bottle. Cross-couponing is often used for the introduction of new products and/ or increase of sales of weak products by linking them with established products in the company's product ling.

Despite all of the above mentioned types/forms of bundling, in practice, companies mostly use the two basic forms of bundling – the pure bundling, which refers to the practice of selling two or more discrete products only as part of a bundle, and the mixed bundling, which refers to the practice of selling a bundle of the products as well as the individual products themselves. The mostly known example of mixed bundling is the McDonald's (Figure 5.7). It is common that when the customer gets to the cashier, he/she is asked whether he/she wants a Combo, or not. The Combo means that the product will include not just the hamburger, but also a medium size drink and fries. This is more price-effective for the consumer compared to buying each of the items individually. Obviously, the company would earn more money this way from three items than to earn from one item alone (Humphries, 2016).

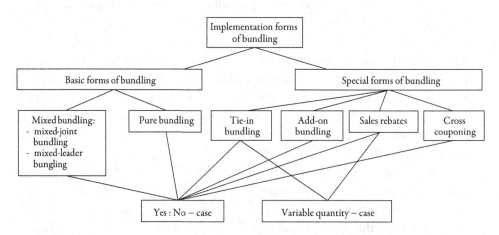

Figure 5.6. Implementation forms of bundling (Fuerderer *et al.*, 2013; pp. 10).

Figure 5.7. Example of bundling at McDonald's.

As another example of mixed bundling can be mentioned the strategy of numerous cinemas, which do not offer only the tickets to the cinema, but also some refreshments, which are offered separately, but also as so called Combo – combination of different products for a better price, e.g. popcorn and drink, nachos with sauce and drink, two drinks for one price, etc. (Figure 5.8). Their aim is to present a bundle of products that seems to offer more than the sum of its parts. 'The sale of the higher priced bundle, instead of just a few of its parts, increases revenue and, as long as it is priced properly, profit' (Beyond Cost Plus, 2016).

Figure 5.8. Example of bundling in cinemas (Beyond Cost Plus, 2016).

Bundling is a strategy, which can be used not just by fast food restaurants and cinemas, but also by restaurants in general, cafeterias, retailers, wholesalers, different supermarkets, hypermarkets and by sellers in general. As examples of bundling in their practice can be mentioned the Outback Steakhouse, or Chili's which have offered a four course menu bundle which included soup, salad, entrée and dessert for $ 15 per person and a promotional deal on social media for $ 35 that included a meal for two (two entrees and Corona-ritas) (Restaurant Engine, 2016); the apples or other types of fruits or vegetables wrapped together in a tray using bopp film; tins wrapped in printed ldpe bundle shrink film using a bundle wrapper; six packs of water, soft drinks, beers, ciders, etc.; multipacks of cereals or sweets and so on.

Box 5.2. Mars customer appreciation day at Wholesale Club Canada (Splash, 2012).

Splash MM recently partnered up with Mars Chocolate in Calgary, Alberta to spread the word about Mars' Customer Appreciation Day which offers some hot new deals in Canada.

Splashers were stationed in Wholesale Clubs across the city where they set-up vibrant, eye-catching displays of Mars products in order to catch the attention of store patrons (Figure 5.9). Splashers engaged shoppers and informed them about the savings and offers available. Retailers were able to save on a buy 10 get on free deal on boxes of Mars Candy, and the public were able to take advantage of buy 8 pouches of M&Ms or Maltesers and get 2 complimentary movie tickets to a Cineplex Odeon near you!

Figure 5.9. Example of bundling at Wholesale Club Canada (Splash, 2012).

Mentioned Combos and special menus are also examples of the so called price bundle, which is characterized as the offer of two or more separate products at a special price, where the consumer can earn few dollars on his purchase. The last very often-used type/form of bundling is so called service, resp. product bundle, which example is the coffeehouse Starbucks, which focuses on providing coffee and tea. Food choices range from pastries and muffins to soups and sandwiches. Starbucks combines them together with the idea of convenience for the office workers and also sell instant coffee in its shops or some supermarkets (Starbucks, 2010).

The primary benefit of product bundling lies in the segmentation ability, by which the consumer excess can be extracted more effectively. (Olderog and Skiera, 2000; Schmalensee, 1984). In marketing, bundling is often described as a form of price discrimination, where we can distinguish between three levels (Pigou, 1920):

1. First-degree price discrimination – also known as perfect price discrimination, where the supplier of some product can sell it to every customer at a different price. The prices for different units of the product can also vary from customer to customer.
2. Second-degree price discrimination – the supplier sells different units of a product for different prices, but every customer who buys the same amount of the product pays the same price. It means that the price depends on the amount of the purchased product.
3. Third-degree price discrimination – occurs when the supplier sells some product to different customer groups or segment at different prices, but every unit of the product sold to a group sells for the same price. This is the most common form of price discrimination, where examples are student discounts, senior discounts, etc. Product bundling is a form of third-degree price discrimination, where the price depends on how much the customers are willing to pay for sold products/services.

5.3 Convenience and bundling trends

As it is clear from the afore mentioned, convenience is multi-faceted and will continue to evolve and converge with other trends to meet the customer's needs and the demands of society, and bundling is a trend that provides added convenience and value to consumers, through reduction of costs of search and related 'transactions' costs. (Faircloth and Connie, 2000) This is also why we are dealing with the convenience and bundling trends together in one chapter. The basic principle of both of them is the consumer's saving of time and effort. In the case of convenience, the savings are due to convenient products, which can save time and effort in preparing and cooking of meals. In the case of bundling the saving is due to offering products in bundles, so that the customer can save not just the time (he does not have to seek for separate products), but also the money (bundled product is in many cases cheaper as the separate products) and the effort (by reducing transaction costs).

Box 5.3. Five tips to drive sales with the art of product bundling (Shelton, 2015).

1. Bundles provide convenience in a fast-moving world. While the philosophy of the old school bundle was rooted in negative synergy (customer would be willing to buy the bundle only if it is cheaper than buying the products individually), the present bundle is driven by positive synergy (customers see the well-arranged bundle as more valuable than its individual components and they are willing to pay for it). The driving value of this purchase is the convenience.

2. Bundling invents new products from old. Bundling allows the seller/retailer to take ordinary items and sell them in a whole new way. Much of the art revolves around intersecting two concepts or creating interesting juxtapositions and bundles are proof that this works in the marketing world.

3. Curation makes customers feel special, creating loyalty. The massive rise in subscription boxes over the last two years has been unprecedented. There are services as for example the Subbly or Cratejoy to help ordinary people curate subscription boxes – each box is mailed regularly to subscribers and contains a small selection of luxury or specialist goods, from fresh gourmet ingredients to beauty products to clothes. There are even specialist boxes such as Lootcrate and Nerdblock, both stuffed with toys and gadgets for a primarily male audience that is notoriously difficult to buy for.

4. Bundles create a gook for new customers. A bundle that a customer feels is particularly valuable, time-sensitive or exciting can provide a great hook for future purchases, introducing them to new products and driving commitment to your brand.

5. Bundles drive your competitive advantage. While the price can be the only distinguishing factor in the online marketplace, in the case of retails, the distinguishing factor can be the bundled product – by the creation of branded gift baskets or repackaging products into an unique selection box the seller can create an individual product, which is difficult for competitors to emulate. The new generation of bundling is a growing art in the marketing world, full of opportunities for the retailers.

References

Adams, W. and Yellen, J., 1976. Commodity bundling and the burden of monopoly. Quarterly Journal of Economics 90: 475-498.

Bae, H.J., Chae, M.J. and Ryu, K., 2010. Consumer behaviors towards ready-to-eat foods based on food-related lifestyles in Korea. Nutrition Research and Practice 4: 332-338.

Berry, T., 2012. Product bundling. Mplans. Product marketing. Available at: http://www.mplans.com/articles/product-bundling.

Beyond Cost Plus, 2016. The appeal of bundled products. Available at: http://tinyurl.com/yagop5zt.

Bodyecology, 2015. The six thousand hidden dangers of processed foods (and what to choose instead). Available at: http://tinyurl.com/yafo5d40.

Boseley, S., 2010. Convenience food changes could save 'thousands of lives'. Food safety. Available at: http://tinyurl.com/yaraw448.

Breidert, C., 2005. Estimation of willingness-to-pay: theory, measurement, application. PhD-thesis, WU Vienna University of Economics and Business. Available at: http://epub.wu.ac.at/1934.

Buckley, M., Cowan, C., McCarthy, M. and O'Sullivan, C., 2005. The convenience consumer and food-related lifestyles in Great Britain. Journal of Food Products Marketing 11(3): 3-25.

Candel, M.J.J.M., 2001. Consumers' convenience orientation towards meal preparation: conceptualisation and measurement. Appetite 36: 15-28.

Cardiff, E., 2013. Convenience foods: not so convenient for your health. One Green Planet. Available at: http://tinyurl.com/hcho9xu.

Cargill, 2017. Convenience food. 2017. Available at: http://tinyurl.com/y92xw488.

Chiambaretto, P. and Dumez, H., 2012. The role of bundling in firms' marketing strategies: a synthesis. Recherche et Applications en Marketing 27(2): 91-106.

Chow, C.K., 2007. Fatty acids in foods and their health implications, 3rd edition. CRC Press, Boca Raton, FL, USA.

Consumer Trend Report, 2010. Convenience. Available at: http://tinyurl.com/ycjwnsev.

Derdenger, T. and Kumar, V., 2013. The dynamic effects of bundling as a product strategy. Marketing Science 32: 827-859.

Faircloth, L. and Connie, M., 2000. Problems surrounding the mortgage origination process: congressional hearing. DIANE Publishing, Washington, DC, USA.

Frewer, L.J., Risvik, E. and Schifferstein, H., 2001. Food, people and society: a European perspective of consumers' food choices. Springer, New York, NY, USA.

Fuerderer, R., Hermann, A. and Wuebker, G., 2013. Optimal bundling: marketing strategies for improving economic performance. Springer, New York, NY, USA.

Guiltinan, J., 1987. The price bundling of services: a normative framework. Journal of Marketing 51(2): 74-85.

Humphries, M., 2016. The game of product bundling: a Nintendo story. Available at: http://tinyurl.com/yd3xxr2z.

Kah wai, 2014. Ways and ideas in advertisement bundle. Available at: http://tinyurl.com/y8mz73jz.

NutritionVista, 2009. Convenience foods – Inconvenient for our health. Available at: http://tinyurl.com/22us743.

Olderog, T. and Skiera, B., 2000. The benefits of bundling strategies. Schmalenbach Business Review 1: 137-160.

Pigou, A.C., 1920. The economics of welfare. Macmillan, London, UK.

Pride, W. and Ferrell, O., 2008. Marketing. South-Western Cengage Learning, Boston, MA, USA.

Recapt, 2011. Retailer and consumer acceptance of promising novel technologies and collaborative innovation management. Available at: http://www.wur.nl/nl/show/recapt.htm.

Rees, J., 2005. Eating properly. Black Rabbit Books, Mankato, MN, USA.

Restaurant Engine, 2016. Bundle your menu items to create appealing 'deals' for customers. Available at: http://restaurantengine.com/bundle-your-menu-items.

Rudolph, T., Schlegelmilch, B., Bauer, A., Franch, J. and Meise, J.N., 2012. Diversity in Europeans: text and cases. Springer, New York, NY, USA.

Schmalensee, R., 1984. Gaussian demand and commodity bundling. Journal of Business 57: 211-230.

Shelton, K., 2015. Retail store trends: product bundling is an art form. ASDinsider. Retailer's Guide. Available at: http://tinyurl.com/ybwgukf3.

Splash, 2012. Mars customer appreciation day at wholesale Club Canada. Available at: http://tinyurl.com/y9wp4clo.

Srinivasan, S. and Shende, K.M., 2014. A study on the benefits of convenience foods to working women. Available at: http://tinyurl.com/y794a3ak.

Starbucks, 2010. Service-product bundle. Available at: http://tinyurl.com/qkw2kcg.

Stremersch, S. and Tellis, G., 2002. Strategic bundling of products and prices: a new synthesis for marketing. Journal of Marketing 66(1): 55-72.

Telser, L., 1979. A theory of monopoly of complementary goods. Journal of Business 52(2): 211-230.

Warde, A., 1999. Convenience food: space and timing. British Food Journal 101: 518-527.

wiseGEEK, 2016. What are convenience foods? Available at: http://tinyurl.com/trn2mn2.

Yadav, M. and Monroe, K., 1993. How buyers perceive savings in a bundle price: an examination of a bundle's transaction value. Journal of Marketing Research 30: 350-358.

Consumer trends and new product opportunities in the food sector

6. Introduction to the food chain

R. Haas* and M. Petz

Institute of Marketing & Innovation, Department of Economic and Social Sciences, University of Natural Resources and Life Sciences, Vienna (BOKU), Feistmantelstr. 4, 1180 Vienna, Austria; rainer.haas@boku.ac.at

Abstract

An overview of the contemporary food chain is given. After a general model, several variations are discussed. An exploration of food chains at different levels – local, regional and global – is covered. These are known as short and long food chains respectively. How consumers make social constructs in mediating the food chain is explored along with a contemporary and futurist perspective on problems to be solved by food chains and related praxis.

Keywords: food value chain, food supply chain, food supply networks

6.1 Introduction

In this chapter we will explain characteristics of food chains and give an overview of different forms of food chains. We will highlight paradoxes of the food chain before we discuss contemporary issues of food chains. At the end of the chapter we look at factors characterizing short (local) vs long (global) food chains.

The production of food leads to a variety of economic, environmental and social implications and connects a set of different stakeholders, namely farmers, distributors, traders, processors, retailers, consumers and disposers. Terms like from 'farm to fork' or from 'stable to table' illustrate that whenever we talk about food chains, we talk about a system linking agricultural producers with consumers. Depending on the number of intermediate facilitators involved in processing and distributing the final food product to the consumer and on the geographical distance between farmer and consumer we can differentiate food supply chains in terms of spatial proximity (short vs long), in terms of business relations of the involved actors, and in terms of form (chain vs network).

Generally, a food supply chain can be defined as 'the set of trading partner relationships and transactions that delivers a food product from producers to consumers' (King *et al.*, 2010). When business relations in the food chain are more transactional than relational – with a focus on the product itself (a tomato is a tomato), and less on the story behind the product or the identity of the producers (origin, production method, philosophy, environmental and social impact, etc.) – the term food supply chain is commonly used. In food supply chains farmers are interchangeable and anonymous. In food value chains farmers and their production method are a crucial part of differentiating the food product from the competition.

We concur with the definition of food value chains (or value-based food supply chains) as 'strategic alliances between farms or ranches and other supply-chain partners that deal in significant volumes of high-quality, differentiated food products and distribute rewards equitably across the chain' (Stevenson and Pirog, 2013). The degree of collaboration is the differentiating characteristic between food chains and food value chains. Food value chains represent business alliances, where the involved actors aim to achieve a competitive advantage by mutual collaboration and accept interdependence as a necessary by-product of their business relations.

Besides the metaphorical chain, networks of collaborating actors are a prominent form in the food sector as well. In food supply networks 'interrelationships evolve dynamically through changing trade relationships from within a network of different enterprises that are active at each one of the stages of the value chain' (Fritz and Schiefer, 2008). In the following we use the term food chain, as it is more generic.

As Figure 6.1 depicts, the first actors in the food chain are input companies. The quality and nature of their inputs is of significant relevance for the rest of the food chain. Similar to eco-systems, once a substance gets introduced into the 'food chain', it travels up the 'food pyramid', accumulating over time in the animals on top of the pyramid, in our case humans. In 2011 Germany faced a dioxin scandal with consequences for the egg- and pig-producing sectors. An input company was accused of having sold feed containing technical industrial fat not approved as animal feed (FAZ.NET, 2011). After the discovery, several countries immediately banned the import of eggs and pig meat from Germany, thus damaging two whole sectors. The importance of the input actors for agriculture and society is exemplified by the ongoing discussion about glyphosate, a substance used in herbicides. Recently, a decision of the EU about the re-approval of glyphosate in all member states has been postponed, due to rising concern over the safety of glyphosate for human health. Glyphosate is under suspicion of being a carcinogen and is furthermore suspected to be an endocrine disrupter (Gasnier *et al.*, 2009; Richard *et al.*, 2005). A study from the Munich Environmental Institute found residues of glyphosate, some to an amount 300 times higher than allowed in drinking water. The German Brewers Association immediately doubted the quality of the study, but admitted at the same time that glyphosate 'is now found virtually everywhere after decades of use in agriculture' (Pesce, 2011). That glyphosate has entered the food chain seems to be confirmed by the results of another study which found glyphosate in blood and urine of Danish dairy cows, which had been fed with GMO corn (genetically modified corn, Krüger *et al.*, 2013). The fact that toxic substances accumulate over time in the food chain and reach the consumers eventually, is the main reason for traceability and quality assurance programs, which demand that farmers document meticulously, which input factors they are using. From a marketing point of view, it is all about establishing trust in the product.

There are many ways how farm products can reach the consumer (Figure 6.1). The most common way is via traders, food processors and then retail. In short supply chains or alternative food networks, farmers sell directly to the consumer. The same goes for food processors. Food processors can either sell their products to retailers or directly to the consumer over e-commerce channels. They produce either their own brands or private labels for a retailer. Private labels were at the beginning mainly positioned as discount products, but nowadays private labels cover the whole price range, many of them offered at premium price levels, especially in the case of organic private label products. Consumers perceive less and less a quality difference between private labels and manufacturing brands, thus increasing the competitive pressure on the food processing sector and their manufacturing brands

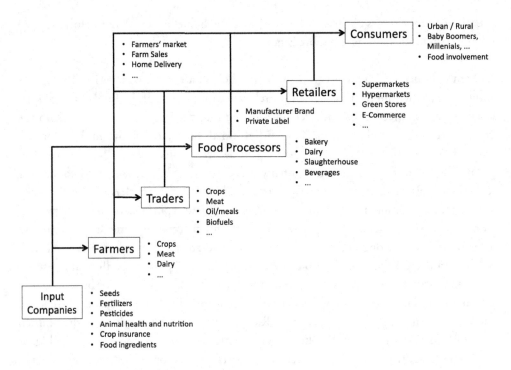

Figure 6.1. Food chain (adapted from Stiring *et al.*, 2013).

(Haas *et al.*, 2012). The increased market share of private labels is an indicator of a shift in market power in the food chain. In the past, the food industry was the dominant actor in the food chain, but with an ever increasing supply of food products, the scarcity factor was no longer the product itself, but the shelf space in the supermarket. Retailers became the most powerful actors in the food supply chain, not only because they were able to limit the access of food companies to their shelves by replacing their brands with private labels, but also by owning highly accurate data about consumer shopping behaviour on a day to day basis. Linking their scanner data with membership and customer loyalty programs allowed them to understand consumer preferences and consumer segments in an unprecedented way, which no other actor in the food chain had been able to achieve. This gives them the ability to stir 'collaborative pull innovation' (Weaver, 2008), by communicating consumer preferences up the food chain. An example is the introduction of strong organic brands all over Europe. Retailers were among the first to introduce and develop premium organic private labels. They clearly understood the strategic impact of offering organic products to improve the image of their supermarkets and to attract wealthy educated consumers (LOHAS – lifestyle of health and sustainability; Haas *et al.*, 2010).

Other examples of the market shaping force of retailers are heirloom variety programs, food waste reduction programs or programs to increase sustainability. An example is the Pro Planet initiative of the German retail group REWE. Pro Planet aims to increase the share of regionally produced food, promotes integrated pest management and runs programs to preserve heirloom fruit and vegetable varieties. As consumers got more aware and concerned about animal welfare, retailers 'motivated' their agricultural producers to increase cage-free egg production. In Austria SPAR and REWE initiated for the same reason breeding programmes of organic chickens, which lay eggs but also produce a sufficient meat quality (Schuh, 2015). The prevalent standard procedure for laying hens is to kill the male chicks right after their birth, also for organic chickens; in Germany alone it is estimated that 50 million male chicks get killed per year (Peta, 2015). The introduction of the so called 'Zwei-Nutzungshuhn' ('dual purpose chicken') shows how consumer preferences are transferred by retailers downstream to farmers resulting in collaborative pull innovation. These examples highlight that powerful actors in the food chain can initiate collaboration beyond company borders and can be a significant source of innovation.

6.2 Paradoxes of the food chain

From a global perspective, the first baffling paradox of food production and consumption is that there are approximately 800 million people suffering from hunger and approx. 1 billion of people suffering from over-nutrition, resulting in overweight and obese people with a rising number of diabetes and cardiovascular diseases. It is estimated that annually 29 million people die worldwide because of over-eating and 39 million die because of hunger (Barilla Center for Food and Nutrition, 2012). The number of people with diabetes more than doubled in three decades from 153 million in 1980 to 347 million in 2008 (Danaei et al., 2011). 70% of this increase is due to population growth and ageing, but 30% of this increase is based on a change towards a modern lifestyle with less physical activity and unhealthy eating habits (Danaei et al., 2011). Childhood obesity is also disturbingly rising. In the USA every fifth child between 2 and 5 years is overweight or obese. And for 2030 the forecast estimates a rise to 472 million people with diabetes worldwide, from whom 80% will be in low or middle income countries, often with no access to insulin drugs and insufficient health care systems (Danaei et al., 2011). For the USA alone the medical cost due to obesity and excess weight gain will lead to additional treatment costs of US$ 48 to 66 billion per year until 2030 due to a rise in cardiovascular diseases, strokes and cancers (i.e. non-communicable diseases, Wang et al., 2011). There are several factors responsible for the rise in obesity and there is no magic bullet to solve the problem (Dobbs et al., 2014), but modern food chains are responsible for the rise of this phenomenon to a certain degree. The food industry and food retail in particular are now accused of being as harmful as the tobacco and alcohol dealers, and the industry is even compared to a disease vector in the highly regarded scientific journal The Lancet: 'In industrial epidemics, the vectors of spread are not biological agents, but transnational corporations. Unlike infectious disease epidemics, however, these corporate disease vectors implement sophisticated campaigns to undermine public health interventions' (Moodie *et al.*, 2013). The ultra-processed food and out of home consumption are important drivers of malnutrition and obesity. Processed food accounts for 75% of world food sales and the largest manufacturers cover one third of the worldwide food market (Moodie *et al.*, 2013).

Some governments have reacted to the obesity problem by imposing a tax on sugar. Mexico, France, Belgium, Hungary and recently the United Kingdom introduced taxes on drinks with added sugar. The FDA in the USA is discussing a new label, which will tell the consumer the amount of added sugar. At the moment, US citizens consume 115 g sugar per day, recommended are 50 g of sugar. The food industry is partly reacting to the concerns over sugar. Nestlé published a sugar policy and by 2015 had already started to reduce sugar content by 10% in a selected group of products (Nordrum, 2015). Kellogg's, PepsiCo and The Coca-Cola Company announced that they are going to reduce the sugar content in some of their products (Sawer, 2016).

Besides the rising market shares of processed foods, portion size has also significantly changed. Portion sizes in fast food restaurants are 2 to 5 times bigger than 20 years ago. In the USA portion sizes are on average 25% bigger than in France (Ledikwe *et al.*, 2005). During the evolution of human beings, we always strived to create food with a higher density of nutrients. So in some way human societies become the victims of their own success. On the other side of the spectrum the nigh on 800 million people suffering from hunger would surely prefer to be victims of such kind of success.

Concerning the phenomenon of hunger, societies have to discuss existing priorities reflected in subsidy and production systems. In other words, as human society we should ask ourselves two questions:
1. Should we produce feed for animals or food for humans?
2. Should we produce fuel for cars or food for humans?

It has to be emphasized that hunger is not due to a lack of productivity, but due to poverty, political instability and missing infrastructure. Concerning productivity we already produce enough food to feed 10 billion people (Holt-Giménez *et al.*, 2012). The majority of agricultural crops produced are either for animal feed or for biofuels. Demands to double food production are based on the paradigm that the rest of the world will also switch to a Western food related lifestyle, with a high share of meat consumption and a significant share of biofuels made from food crops. Globally one third of crop production is used to feed animals, in highly developed countries even two thirds of grain are used for animal feed. In the USA 50% of the corn harvest goes into feed lots and 45% is used to produce ethanol for fuel (Barilla Center for Food and Nutrition, 2012). The rise of a wealthy middle class in China, Russia, Brazil and India goes hand in hand with a rise in meat consumption (in India with a focus on chickens). It is a realistic scenario that meat consumption will continue to increase even when Europe and the USA have already reached their meat peak. 'The call to double food production by 2050 only applies if we continue to prioritize the growing population of livestock and automobiles over hungry people' (Holt-Giménez *et al.*, 2012). So the question whether we are able to feed a population of 9 billion people in 2050 is strongly related to future food consumption habits and the share of biofuels produced with food crops.

Meat production demands the most land and water resources compared to crops, fruits or vegetables. A third wave of outsourcing called land grabbing is a phenomenon that emerged around the turn of the millennium and got momentum after the financial crisis of 2008. Land grabbing 'refers to the purchase or lease of vast areas of land by wealthier, food insecure nations and private investors from mostly poor, developing countries with a view to producing crops for export' (Zolin and Braggion, 2013). But not only developing countries are in need of additional land resources. Europe is the biggest importer of food worldwide. A study from the Humboldt University in Berlin estimated the land size necessary to produce the amount of imported crops to the EU, the so-called 'virtual net land import', which is 35 million hectares, an area almost the size of Germany (Von Witzke and Noleppa, 2010). 50% of this virtual Germany is covered by soybeans, most of it GMO soybean, which are used for livestock rearing in the EU, going hand in hand with negative externalities such as reduction of natural habitats with Land-Use and Land Cover Change (LULCC). Deforestation of tropical rain forests and subsequent conversion into crop- or range-land with associated increasing greenhouse gas emissions illustrate this. The fact that 80% of the protein requirements for raising livestock in the EU is imported from non-EU countries, highlights the importance of EU research projects like PROteINSECT (http://www.proteinsect.eu) about the safety of insects as a novel source for proteins in animal feed. Due to the emergence of mad cow disease in the 1980s in the United Kingdom, legislation in the EU forbids the use of animals (also insects) as feed for animals. If research projects like PROteINSECT could prove the safety of insects to be used as animal feed, a change in legislation would be the next step to establish a new market to produce animal feed, rich in proteins, thus diminishing the environmental burden on producing plant based proteins.

6.3 Market-related issues of food chains

As we already mentioned in the introduction, modern food chains are characterized by a high concentration of retail companies, while the food processing sector mainly consists of small and medium sized enterprises (SMEs). In Germany, Edeka, REWE, Aldi and the Schwarz Group (Lidl and Kaufland) comprise 85% of the market (Bielefeld, 2014), in the UK the four top retailers cover 70% of the retail market (Statista, 2016) and in France the top three retailers cover almost 50% (Statista, 2013).

The competition of food companies is shifting towards a competition of food chains and food networks with varying degrees of coordination between the network members. As a result, product differentiation at the consumer level goes hand in hand with chain differentiation. The competing food chains are increasingly differentiated: ranging from discount supermarket chains (Aldi, Lidl, Walmart, Trader Joe's) via medium price level supermarket chains (Tengelmann, Carrefour, Tesco, Billa, Merkur) to premium outlets (organic supermarket chains like Denn's or Basic in Germany and Fresh Market or Whole Foods in the USA). In response to the rising income and asset disparities in Europe and the USA, retailers expand their discount and premium product segments, not only by offering low price and premium manufacturer brands, but also by their own discount and premium private labels (Haas *et al.,* 2012).

The innovations in IT (information technology) in recent decades changed the face of retail significantly. Barcodes, scanner data, web-based EDI (Electronic Data Interface) have provided retailers with instant access to daily sales patterns and changes in consumer preferences. The establishment of e-procurement systems and web-based EDIs improved the information exchange between suppliers and retailers and lowered the transaction costs. The emergence of collaborative planning, forecasting and replenishment reduced storage costs and increased the accuracy of sales forecast. The formation of Efficient Consumer Response, a joint retail and industry organization, exemplifies how actors in the food chain collaborate to make the sector more responsive to consumer demands and to further optimize costs in the supply chain.

Selling food over e-commerce websites shows worldwide promising growth rates, but the market shares are still marginal compared to conventional distribution channels. In 2014, the share of online sales for groceries and drugstore products, in Germany was 1.2% compared to food sales over conventional retail. This represents a turnover of € 2.6 billion in Germany (GfK, 2015). GfK forecasts an increase of 400% for Germany until 2025, but yet that would only represent 5% of total food sales. In Austria, the share of food e-commerce compared to total food retail sales is also 1%, compared to computer electronics it is 21% (KMU Forschung Austria 2014). In the USA it is even less than 1%, but forecasts estimate a compound annual growth rate of 21% until 2018, which significantly surpasses growth rates of traditional supermarkets (Sherbrooke Capital, 2016).

Important strategic and operational goals of food chains are resource efficiency (optimization of logistics, reduction of food waste), food safety, transparency and sustainability. A survey of the *Lebensmittel Zeitung* trade publication asked consumers about the top ten trust factors with respect to food producers. Four of ten factors were related to transparency (Table 6.1).

Transparency is an important means to establish trust. Consumers are increasingly demanding to know where the ingredients in their food came from, or if ethical standards have been applied in its production and sourcing. An example of how to establish transparency is the award winning campaign from McDonalds 'Our Food your Questions', which originally started in Canada, but got later rolled-out to other countries. Consumers are invited to ask questions and the answers are posted on a dedicated website (Birkner, 2014).

Due to several food scandals over the past few decades (antibiotic-resistance, steroid hormones in feed, dioxin poultry affair, horse-meat scandal, melamine in milk) governments introduced legislation for stricter traceability. Food traceability became a cornerstone of EU food safety policy. The EU quality system is a comprehensive, integrative approach ('from stable to table'). Agriculture and food and feed processing companies are primarily responsible, that food and the components of food are traceable, and where a uniform, comprehensive risk analysis and risk assessment system guarantees highest food quality

Table 6.1. Top 10 trust factors in food producers (Lebensmittel Zeitung/Musiol Munzinger Sasserat, 2012); n=1000).

1.	Quality of products and services
2.	Transparency of ingredients
3.	Transparency of origin
4.	Transparency of companies to handle food scandals or other problems
5.	Reliability of food company
6.	Warranties on products and offers
7.	Transparency of production
8.	Comprehensible price formation
9.	Labelling of food products
10.	Social fairness in respect to employees

within all stages of the supply chain. The basic EU risk alert system is called RASFF, the Rapid Alert System for Food and Feed. The system was introduced more than 35 years ago and led to the EU having one of the highest food safety standards in the world. This is mainly due to the solid set of EU legislation in place, which ensures that food is safe for consumers. Consumers can trust in the food and feed within the EU market. RASFF is a key tool to ensure the cross-border follow up of information to swiftly react when risks to public health are detected in the food chain. Important here is the concept of food integrity (sometimes termed food authenticity), which attempts to deal with food fraud.

> Food fraud is a collective term used to encompass the deliberate and intentional substitution, addition, tampering, or misrepresentation of food, food ingredients, or food packaging; or false or misleading statements made about a product, for economic gain
>
> (Spink and Moyer, 2011)

Transparency and sustainability became core areas for the food chain, and companies are advised to take their efforts beyond regulatory requirements. The fact that sustainability (food waste reduction programmes are a part of it) is of increasing importance in the food chain is on one side a direct response to changing consumer demands and on the other side a move towards resource efficiency. The already mentioned ProPlanet initiative of REWE (http://www.proplanet-label.com) or the Billa sustainability program in Austria ('Wer nicht von gestern ist, denkt an morgen!' – i.e. Who is not from yesterday, thinks about tomorrow!) are indicators that retailers are taking this issue seriously. However, retailers alone would not be able to establish sustainability in the food chain without including other major stakeholders such as the food manufacturing industry, government institutions and/or farmers. A study about corporate social responsibility (CSR) in the food chain identified 216 CSR standards, guidelines and initiatives in the agribusiness sector (Poetz *et al.*, 2012, 2013). Some were founded by consortia of the food industry such as SAI (Sustainable Agriculture Initiative), which was established by Groupe Danone, Nestlé and Unilever, but most of the CSR schemes claim to be based on multi-stakeholder involvement. The last two decades saw an emergence of numerous CSR schemes, demonstrating that sustainability has become a serious item on the public and corporate agendas.

'Companies have been increasingly offered guidance on how to get away from a single bottom line and realize social and environmental responsibility both in and beyond compliance with existing laws. Furthermore, standards and guidelines on how to credibly account for CSR measures and report on the progress have been developed. Published October 2010, the much anticipated ISO 26000 Guideline on Social Responsibility represents the peak of the proliferation of CSR standards' (Poetz *et al.*, 2013).

It is no surprise that the majority of CSR schemes addresses specifically one product or one food chain. There are CSR schemes for cocoa, tea and coffee (UTZ, Starbucks Coffee), palm oil (RSPO – Roundtable on Sustainable Palm Oil), soybean (RTRS – Round Table on Responsible Soy), milk and dairy products (Mother Dairy), fish (MSC – Marine Stewardship Council), wine, flowers (Fairtrade), cotton, sugar, sugarcane, fruit processing and tobacco (Poetz *et al.*, 2013). The Rainforest Alliance Sustainable Agriculture Network (SAN) Standard starts at the farm level and includes commodities such as coffee and bananas, sugarcane, palm oil, sunflower, soybean and peanuts (SAN 2016). Often CSR schemes emerge as a response to increased public scrutiny, for example those on soybean or palm oil in response to criticism about deforestation and loss of biodiversity.

Reducing food waste is another important issue for the food chain, especially with respect to the growing world population. Europeans and North Americans throw on average between 95 and 115 kg per capita/per annum of food away (Gustavsson *et al.*, 2011). In general, 30-40% of food is wasted in developing and developed countries, but the areas where food waste occurs are different (Figure 6.2). The majority of food waste in developing countries occurs mainly at the farm, transporting and processing levels, due to lack of infrastructure in the food chain (cold storage facilities) and absence of investment and knowledge of storage facilities at farm level (Godfray *et al.*, 2010). India alone wastes 30 to 40% of fresh fruits and vegetables, because of a lack of cold storage at wholesaler and retailer level. In Southeast Asia 30% of rice is lost post-harvest due to pests and spoilage (Godfray *et al.*, 2010).

In developed countries, food waste on the farm is much less of an issue, the biggest losses occur at the retail, food service and domestic links of the chain. The reasons are manifold, food is relatively cheap in relation to household income, and many consumers don't understand the safety margins of 'best before' dates and their difference from the 'use by' expiry dates. Additionally, legislation does not allow re-use of food containing animal proteins as animal feed, because of control of mad cow disease. The EU regulation 1774/2002 defines three categories of animal by-products. Only category 3, which contains expired food products, which are of animal origin or contain ingredients of animal origin and were produced for human consumption, are allowed to be used for pet food. Category one and two have to be either incinerated or composted. Allowed is the use of dairy products and eggs for animal feed, if no meat products are contained. In the past, feeding cooked food waste – also known as swill – to pigs was a common practice. Since the outbreak of foot and mouth disease in the United Kingdom it was banned in 2002 in the EU as animal feed. In many countries outside the EU the use of swill as pork feed is allowed (e.g. Thailand, South Korea, Japan). A recent study estimates that the allowance of swill as pork feed could reduce the land use of EU pork by 20% and save 1,8 Mio. hectares (Zu Ermgassen *et al.*, 2016).

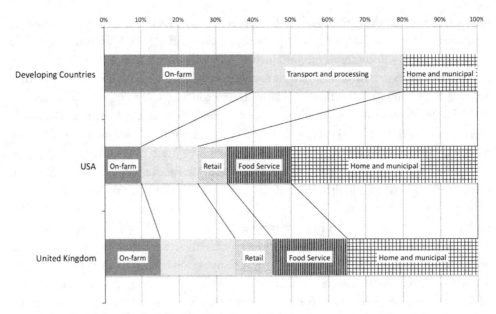

Figure 6.2. Total food waste in developed and developing countries. Retail, food service, home and municipal are grouped together for developing countries (Godfray *et al.*, 2010).

The use of expired food products is defined by EU regulations 197/2006, 832/2007 and 129/2009. Expired food products which did not get in contact with products of animal origin and that contain no meat and meat products are allowed to be used for animal feed without further treatment, an example are bread and pastries (Schneider and Scherhaufer, 2009).

In attempts to deal with food waste in a responsible way we see the emergence of initiatives collecting surplus food from supermarkets and sharing it with charities, providing food to low income groups. Foodsharing.de is an initiative in Germany organized by consumers who are willing to share food they have in excess with other consumers. In Germany, most retailers cooperate with 900 initiatives called 'Die Tafel' to provide food still eligible for human consumption to low income groups. France released a law that forbids supermarkets to throw food away. Since July 2016 all supermarket chains in France with outlets bigger than 400 m2 have to collaborate with charity organisations. Social media play an important role in that field in connecting consumers engaged in food sharing. The most extreme form of collecting discarded food is probably dumpster diving, where people who could afford to buy food, climb into garbage dumpsters to collect what has been thrown away. Often the dumpster dived food is reused in collective kitchens.

6.4 Global food chains versus short local food chains

It is important to note that there is no such thing as one commonly shaped food chain. Globalization has promoted the emergence of competing global food chains. The resulting complex global food chains span long distances from production of raw materials, trade, transport, processing, and finally consumption. In a typical hypermarket consumers face between 25,000 and 40,000 products (FMI, 2011). In the USA the majority of food products travelled 2,500 to 4,000 kilometres from farm to the supermarket shelves (Worldwatch Institute, 2008) and 85% have their origin in other states or countries. As a response to the increased global and anonymous product assortment, we see a counter grass root movement of short food chains (also called alternative food networks) producing local food or regional food.

> A key characteristic of short supply chains is their capacity to re-socialize or re-spatialize food, thereby allowing the consumer to make value-judgements about the relative desirability of foods on the basis of their own knowledge, experience, or perceived imagery. Commonly such foods are defined either by the locality or even the specific farm where they are produced; and they serve to draw upon and enhance an image of the farm and/or region as a source of quality foods. 'Short' supply chains seek to redefine the producer-consumer relation by giving clear signals as to the origin of the food product
> (Marsden *et al.*, 2000)

An innovative example illustrating the grass roots nature of short food chains and how they redefine the producer-consumer relation is the Food Assembly Network (https://thefoodassembly.com), which was founded in France, but afterwards spread to Germany. It is a web-based platform, which uses the internet to bring local producers together with consumers in cities. Every 'cell' of the Food Assembly Network contains a list of local producers from which consumers can order their product. Once a week farmers deliver to a store in a city, where the consumers go to pick up their orders ('click-and-collect'-service). Other forms of alternative food networks, which build on the consumer desire to establish a relationship with farmers or to be involved in producing their own food, are community supported agriculture, community gardens, urban gardening or the rising trend of foraging.

From the consumer perspective, long complex food chains often represent a black box, making it almost impossible to know where the raw material comes from and how it has been produced. Many consumers are not aware or do not care about how their daily food choices are connected to environmental (palm oil and extinction of orang-utans in Indonesia) and social issues (cocoa and child labour) in regions far away from their buying decision. But for more and more consumers ethical and environmental aspects are increasingly important, especially for the generations of Baby Boomers and Millennials compared to the post-war generation. This shift in consumer preferences results in a communication down the food value chain to the farmers. In response famers and food processors adapt new product features such as social responsibility and environmental friendliness. Global supply chains respond by increasing their transparency and offering new production standards (see CSR in the text above), and short food supply chains profit from these new demands.

Despite the fact that consumers define local or regional food very differently – ranging from 50 km to the whole country – they apparently use the attribute local/regional as a quality cue for food products (Cerjak et al., 2014; Haas et al., 2013). 'A common characteristic, however, is the emphasis upon the type of relationship between the producer and the consumer in these supply chains, and the role of this relationship in constructing value and meaning, rather than solely the type of product itself' (Marsden et al., 2000). The main motivations for consumers to buy local food are to promote health and to support the local community, so that food is used to create identity and a feeling of belonging. Interestingly, when asked about product attributes of local food, consumers respond with 'it is healthier', 'fresher', 'natural production/protect the environment' and 'tastier' (Haas et al., 2013). The fact that there is no scientific proof that local/regional food is healthier, or that natural/local production is not in all cases more environmental friendly (Schönhart et al., 2009), reveals that consumers apply quality cues for a variety of attributes which do not necessarily need to have a scientific basis.

Nevertheless, short supply chains play an important role for rural livelihoods. Especially for small and medium sized farms, which often struggle to achieve economies of scale. Short supply chains offer opportunities to achieve higher price margins for farm products, consecutively generating higher income compared to conventional supply chains. '... for a farm of my size I can secure a premium worth an extra £5,000. That is a significant mark-up on the price that could be secured elsewhere' (Marsden et al., 2000). In 2013 for each food dollar spent, farmers in the USA received 17.4 cents (USDA, 2015). A study from the USDA Economic Research Service compared five products from local food chains with national food chains and found that the net revenue of farmers is equal or up to seven times higher than in long food chains (King et al., 2010).

There are three forms of short food chains (Marsden et al., 2000):
1. face-to-face;
2. spatial proximity;
3. spatially extended.

In face-to-face food chains consumers buy directly from the farmer/processor. Personal face-to-face communication is the basis to establish trust and authenticity. This is what generally would be called 'local food'. The main quality cue is the spatial proximity, which generates trust. It is not the name of the region, simply because it has not been established as a 'brand' yet (like Champagne or Parma).

Short food chains with spatial proximity produce and sell their products in the same specific region. Consumers are informed about the local food product at the point of sale. Examples would be the brand 'Roter Hahn' in South Tyrol, which sells regional specialties from South Tyrol mainly to tourists visiting South Tyrol, or the 'Genuss Region Österreich' (i.e. Austrian Region of Delight). This brand stands for regional, culinary specialties from selected regions of Austria (Figure 6.3). Regional products like these examples combine specific authentic food products, embedded in a beautiful cultural landscape, with traditional artisanal production. Consumers can buy these products in local supermarkets, specialty stores as well as in restaurants.

Spatially extended short food chains sell locally produced food products to consumers, who live outside of the region of production. The European label of protected designation of origin (PDO) or protected geographical indication (PGI) are typical examples of regional food products falling into this category. Other typical examples of regional products would be Champagne, Parmigiano-Reggiano or Kalamata Olives.

Those three different short food chains also implicitly reflect the evolutionary aspect of short food chains. Short food chains start face-to-face and their ideal way to develop is towards spatially extended short food chains. The potential of origin-linked products depends on

Figure 6.3. Examples of Genuss Region Österreich Label (Austrian Region of Delight).

the territorial capacity of the region they come from. Territorial capacity encompasses the traditional production method, the quality of the raw material, the artisanal knowledge of local processing companies, the specific merits of the local climate and soil, and finally the beauty of the landscape (Vandecandelaere *et al.*, 2009).

6.5 Conclusion

Food chains are complex dynamic systems, which can be differentiated by the nature of collaboration between stakeholders (transactional vs relational), by spatial proximity (local vs global) or by form (chain vs network). Food chains can have differing strategic orientations, ranging from cost efficiency, value creation, resource efficiency, environmental friendliness or social justice. Food chains can be a mean to stimulate rural development and definitely play a role in fighting hunger and moving wealth between sectors. The constantly growing world population, the rise of non-communicable diseases such as diabetes, the multitude of environmental and social consequences of specific production methods, set the framework for a future in which food chains will have to find better answers, in how to fulfil their responsibility towards society and shareholders.

References

Barilla Center for Food and Nutrition, 2012. Double pyramid 2012: enabling sustainable food choices. Available at: http://tinyurl.com/ycsaernd.

Bielefeld, H., 2014. Konzentration im Einzelhandel hat kaum negative Folgen. Lebensmittel-zeitung. Available at: http://tinyurl.com/y6wxao2l.

Birkner, C., 2014. A peek behind the arches. Marketing News January: 45-50.

Cerjak, M., Haas, R., Brunner, F. and Tomić, M., 2014. What motivates consumers to buy traditional food products? Evidence from Croatia and Austria using word association and laddering interviews. British Food Journal 116: 1726-1747.

Danaei, G., Finucane, M.M., Lu, Y., Singh, G.M., Cowan, M.J., Paciorek, C.J., Lin, J.K., Farzadfar, F.F., Khang, Y.H., Stevens, G., Rao, M.F., Ali, M.K., Riley, L.M., Robinson, C.A. and Ezzati, M., 2011. National, regional, and global trends in fasting plasma glucose and diabetes prevalence since 1980: systematic analysis of health examination surveys and epidemiological studies with 370 country-years and 2.7 million participant. Lancet 378: 31-40.

Dobbs, R., Sawers, C., Thompson, F., Manyika, J., Woetzel, J., Child, P., McKenna, S. and Spatharou, A., 2014. Overcoming obesity: an initial economic analysis. McKinsey Global Institute, Washington, DC, USA, pp. 120.

FAZ.NET, 2011. Überhöhte Giftwerte auch bei Fleisch. Available at: http://tinyurl.com/yb5j3v6k.

FMI, 2011. Supermarket facts. Industry overview 2011. Available at: http://www.fmi.org/research-resources/supermarket-facts.

Fritz, M. and Schiefer, G., 2008. Innovation and system dynamics in food networks. Agribusiness 24: 301-305.

Gasnier, C., Dumont, C., Benachour, N., Clair, E., Chagnon, M.C. and Séralini, G.E., 2009. Glyphosate-based herbicides are toxic and endocrine disruptors in human cell lines. Toxicology 262: 184-191.

GfK, 2015. eCommerce: Germany's online share of groceries and drugstore items to double. Available at: http://tinyurl.com/yatr4h44.

Godfray, H.C.J., Beddington, J.R., Crute, I.R., Haddad, L., Lawrence, D., Muir, J.F., Pretty, J., Robinson, S., Thomas, S.M. and Toulmin, C., 2010. Food security: the challenge of feeding 9 billion people. Science 327: 812-818.

Gustavsson, J., Cederberg, C., Sonesson, U., Otterdijk, R. and Meybeck, A., 2011. Global food losses and food waste. extent, causes and prevention. Food and Agriculture Organization of the United Nations, Rome, Italy, 37 pp.

Haas, R., Canavari, M., Pöchtrager, S., Centonze, R. and Nigro, G., 2010. Organic food in the European Union: a marketing analysis. In: Haas, R., Canavari, M., Slee, B., Tong, C. and Anurugsa, B. (eds.). Looking East looking West organic and quality food marketing in Asia and Europe. Wageningen Academic Publishers, Wageningen, the Netherlands, pp. 21-46.

Haas, R., Sterns, J., Meixner, O., Nyob, D.-I. and Traar, V., 2013. Do US consumers' perceive local and organic food differently? An analysis based on means-end chain analysis and word association. International Journal of Food System Dynamics 4: 214-226.

Haas, R., Weaver, R. and Pöchtrager, S., 2012. Private labels: a strategic tool for retailers? In: Briz, J. and De Felipe, I. (eds.) Food value chain networks in the 21st century. Editoral Agricola Espanola, Madrid, Spain, pp. 225-252.

Holt-Giménez, E., Shattuck, A., Altieri, M., Herren, H. and Gliessman, S., 2012. We already grow enough food for 10 billion people and still can't end hunger. Journal of Sustainable Agriculture 36: 595-598.

King, R., Hand, M., DiGiacomo, G., Clancy, K., Gomez, M., Hardesty, S., Lev, L. and McLaughlin, E., 2010. Comparing the structure, size, and performance of local and mainstream food supply chains. Economic Research Report 81, U.S. Department of Agriculture, Economic Research Service, Washington, DC, USA.

KMU Forschung Austria, 2014. Internet-Einzelhandel 2014. Available at: http://tinyurl.com/y8g3w6s2.

Krüger, M., Schrödl, W., Neuhaus, J. and Shehata, A.A., 2013. Field investigations of glyphosate in urine of Danish dairy cows. Journal of Environmental and Analytical Toxicology 3: 100-186.

Lebensmittel Zeitung and Musiol Munzinger Sasserath GmbH, 2012. Markenvertrauen: eine Bevölkerungsrepräsentative Studie von Lebensmittel Zeitung und Musiol Munzinger Sasserath. Available at: http://tinyurl.com/ybkdh23t.

Ledikwe, J.H., Ello-Martin, J.A. and Rolls, B.J., 2005. Modifying the food environment: energy density, food costs, and portion size portion sizes and the obesity epidemic. Journal of Nutrition 135: 905-909.

Marsden, T., Banks, J. and Bristow, G., 2000. Exploring their role in rural development food supply chain approaches. Sociologia Ruralis 40: 424-438.

Moodie, R., Stuckler, D., Monteiro, C., Sheron, N., Neal, B., Thamarangsi, T., Lincoln, P. and Casswell, S., 2013. Profits and pandemics: prevention of harmful effects of tobacco, alcohol, and ultra-processed food and drink industries. Lancet 381: 670-679.

Nordrum, A., 2015. FDA is cracking down on added sugars: how will big food companies respond? International Business Times. Available at: http://tinyurl.com/y8syjvsc.

Pesce, N.L., 2011. Popular German beers contain weed-killer linked to cancer, study finds. Available at: http://tinyurl.com/hxhz5ry.

Peta., 2015. 50 Millionen Eintagsküken landen in Deutschland jährlich auf dem Müll. Available at: http://tinyurl.com/yb2dgqgb.

Poetz, K., Haas, R. and Balzarova, M., 2012. Emerging strategic corporate social responsibility partnership initiatives in agribusiness: the case of the sustainable agriculture initiative. Journal on Chain and Network Science 12: 151-165.

Poetz, K., Haas, R. and Balzarova, M., 2013. CSR schemes in agribusiness: opening the black box. British Food Journal 115: 47-74.

Richard, S., Moslemi, S., Sipahutar, H., Benachour, N. and Seralini, G.E., 2005. Differential effects of glyphosate and roundup on human placental cells and aromatase. Environmental Health Perspectives 113: 716-720.

Sawer, P., 2016. Soft drink giants to cut sugar levels – But Coke stays the same. The Telegraph. Available at: http://tinyurl.com/ycbk3ey2.

Schneider, F. and Scherhaufer, S., 2009. Aufkommen und Verwertung ehemaliger Lebensmittel – Am Beispiel von Brot und Gebäck. Bundesministerium für Wirtschaft, Familie und Jugend, Vienna, Austria, 103 pp.

Schönhart, M., Penker, M. and Schmid, E., 2009. Sustainable local food production and consumption: challenges for implementation and research. Outlook on Agriculture 38: 175-182.

Schuh, K., 2015. Bioeier: Das Ende der Eintagsküken. Die Presse. Available at: http://tinyurl.com/y7df03v2.

Sherbrooke Capital, 2016. Groceries are the biggest untapped opportunity in e-commerce. Available at: http://tinyurl.com/y93yhleu.

Spink, J. and Moyer, D.C., 2011. Defining the public health threat of food fraud. Journal of Food Science 76: R157-R163.

Statista., 2013. Leading three grocery retailers in France in 2013, by food retail format market share. Available at: http://tinyurl.com/qrdfhka.

Statista, 2016. Market share of grocery stores in Great Britain for the 12 weeks ending January 31, 2016. Available at: http://tinyurl.com/hgwlllf.

Stevenson, S. and Pirog, R., 2013. Values-based food supply chains: strategies for agri-food enterprises-of-the-middle. Available at: http://tinyurl.com/ycpezx02.

Stiring, C., Kruh, W., Proudfood, I., Claydon, L. and Stott, C., 2013. The agricultural and food value chain: entering a new era of cooperation. Available at: https://tinyurl.com/y8v759cn.

United States Department of Agriculture (USDA), 2015. Food dollar series. Available at: https://tinyurl.com/y7s2culr.

Vandecandelaere, E., Arfini, F., Belletti, G. and Marescotti, A., 2009. Linking people, places and products, a guide for promoting quality linked to geographical origin and sustainable geographical indications. Food and Agriculture Organization of the United Nations, Rome, Italy. Available at: https://tinyurl.com/reurz57.

Von Witzke, H. and Noleppa, S., 2010. EU agricultural production and trade: can more efficiency prevent increasing 'land-grabbing' outside of Europe? Available at: https://tinyurl.com/vbzvnlp.

Wang, Y.C., McPherson, K., Marsh, T., Gortmaker, S.L. and Brown, M., 2011. Health and economic burden of the projected obesity trends in the USA and the UK. Lancet 378: 815-825.

Weaver, R.D., 2008. Collaborative pull innovation: origins and adoption in the new economy. Agribusiness 24: 388-402.

Worldwatch Institute, 2008. State of the world: innovations for a sustainable economy. (L. Starke, ed.). W.W. Norton & Company, Washington, DC, USA. Available at: http://tinyurl.com/yaxz9xdd.

Zolin, M.B. and Braggion, M., 2013. Land grabbing, food security and energy security in Asia: the cases of China and India. Proceedings of 3rd Asia-Pacific Business Research Conference. February 25-26, 2013. Kuala Lumpur, Malaysia, pp. 1-13.

Zu Ermgassen, E.K.H.J., Phalan, B., Green, R.E. and Balmford, A., 2016. Reducing the land use of EU pork production: where there's swill, there's a way. Food Policy 58: 35-48.

7. New developments in the farming sector

D. Brohm* and N. Domurath

INTEGAR GmbH, Schlüterstr. 29, 01277 Dresden, Germany; brohm@integar.de

Abstract

The future of agriculture and horticulture just has begun. But many developments are unknown to consumers. One reason might be that people don't connect farming with high-tech. They rather think food comes from an old fashioned farmer or a gardener with a straw hat. On the other side, a majority still believes farming is a job for simple labour. Actually agriculture is one of the less popular fields of study. This situation leads to a deficit of well-trained junior staff, which is one reason for the agriculture industry to develop new techniques to grow and breed more efficiently. Further factors are the limitation of area, a higher food demand, the high usage of exhaustible raw materials like fertiliser, the shortage of fresh water and many more. This chapter describes a couple of up-to-date technologies, prospective developments and visions of high-tech agriculture. Self-driving field vehicles nowadays are in the position to plan their trips alone and even to react to sudden changes. Applications like sawing, spraying, fertilising and harvesting become more popular. The job of future farmers is more a job at a computer screen than on the field. Self-flying drones support their work with graphical material about the status of ripeness, pests and diseases, damages and more. Drones could even be used for activities like spraying. In the field of agriculture and horticulture robots will be an inherent part of the production. Cutting, feeding and milking are current developments. Spraying, harvesting and weeding will be short term technologies. Besides technological development, biotechnology will help to manage the challenges of the future. Genetically modified organisms shall contribute, at least in part, to solute food problems. People more and more live in cities. Bringing food production into urban areas and developing vertical farms is maybe the most popular but still challenging idea.

Keywords: driverless, drones, robots, vertical, biotechnology

Klaus G. Grunert (ed.) **Consumer trends and new product opportunities in the food sector**
DOI 10.3920/978-90-8686-852-0_7, © Wageningen Academic Publishers 2017

7.1 Driverless field vehicles

Self-driving cars are currently in the focus of the car industry worldwide, but it will take another decade to realise them as a ready-to-use solution. However, driverless field machines have been real for a decade, but it took a long time to become real. First attempts were undertaken in the late 1930s by an American farmer. To guide his driverless tractor, a barrel or fixed wheel would be put in the centre of the field and around it would wind a cable attached to a steering arm on the front of the tractor (Condon and Windsor, 1940). A typical round shaped field was the result. The reason for this invention was the long and boring time the farmer used to need before to only drive the tractor. This simple technology was useful but not consistent. The idea of round fields still exists only for the centre pivot irrigation. Here, a long irrigation pipe with sprinklers on wheels is fixed in the middle of the field and turns around the centre (Mader, 2010).

There were no major advances in driverless technology until the middle of the 1990s. Engineers at a British research institute developed a picture analysis based system to guide a small tractor on a field (Williams, 2002). At the same time the Global Positioning System (GPS) was unlocked by the U.S. government for civil applications, which led to a lot more possibilities in the navigation of vehicles. On the one hand there are major differences between cars and trucks driving on a street and vehicles performing on a field. On the other hand both types have to follow fundamental rules and algorithms (Table 7.1).

In the beginning GPS was used for so-called precision farming or precision agriculture applications. Here a driver only supervises the tractor in cases of transposition manoeuvres at the end of the field, in emergency cases or if an impediment e.g. an animal, suddenly appears. The tractor just follows a prepared route across the field. But precision farming is primarily a farming management concept. The base here is the greatest possible knowledge of the field and plants. This information is collected by a couple of sensors before or during the field

Table 7.1. Comparison of self-driving vehicles on the field and on the road.

	Self-driving field vehicles	Self-driving cars
Unpredictable ground	Yes	Rather improbably
Sudden impediments	Reaction is required	Reaction is required
Task	An area has to be processed in the most efficient way	A route from A to B has to be driven in the most fast/ economical/ecological/safe way
Speed	Slow to middle	Slow to very high

work and is analysed and connected to the exact field positions afterwards. So a map of the whole area with the states and demands is generated. The idea behind is to get maximum crop output by minimum but target-oriented input of resources. Irrigation, fertilisation, plant protection and other treatments can be applied on exact demand. Some applications actually must be super precise. In the horticulture sector sometimes it is necessary to meet accuracy of 1 cm, e.g. for making dams or harvesting little crops. GPS exactness actually is only 1 to 50 meters, caused by atmospheric variations of the satellite signals. Here the use of a stationary reference antenna helps to reduce the inaccuracy (Murray, 2008). The high exactness of precision farming applications is only possible in combination with autonomously moving field vehicles and cannot be operated by humans, even if a supervising person might be on board already for emergencies.

After supervised self-drivers the next step in the development was the so called Follow-me technology. Here only the first vehicle in the line is manned. All the others, e.g. tractors, trailers or harvester, follow behind, with the same speed. In South America up to twenty soybean harvesters are led by only one supervised self-driven vehicle. Field machine companies are working hard on making human attendance unnecessary on the field. But supervising the machines will always be required, either on the field or in the office.

The major advantages of self-driving vehicles are:
▶ working around the clock – continuous field work is possible, even at night;
▶ no human failures – sensors never fell asleep or get inattentive;
▶ less labour costs – self-driving vehicles are more expensive, but save a lot of labour costs.

The future of self-driving vehicles could be imagined as fully automated crop management systems. All activities from the field preparation via sawing, spraying, fertilising till the harvest will automatically be scheduled by computer applications only supervised by a highly qualified office farmer. Tractors and accessory equipment not only drive autonomously, they'll also maintenance themselves. Refuelling, loading and unloading, cleaning and even repairing could be expected tasks of future field vehicles.

7.2 Drone technology

Unmanned aerial vehicles (UVA, syn. drones) have been developed by the military to avoid a loss of human life during hazardous manoeuvres. First radio remote-controlled aircrafts were constructed in the early 1930s by the British Royal Air Force. Since then countless types of UVAs in any size for very different applications have been developed. Basically, two types of drones can be distinguished. One type is controlled by remote control of a pilot on the ground or in another vehicle. Other types fly autonomously on base of on-board computer programs. One of the most interesting trends in hobby applications is the use of drones for video or photography. During recent years performance and operating distance have highly increased, especially in the field of multicopters. Even heavy cameras can be placed without problems. But how can we use UVAs in agriculture? Using drones is meaningful if tasks are dull, dirty or dangerous (Tice, 1991). Monitoring is a big issue in plant and animal production. Crops can be controlled with regard to pests and diseases, ripeness, nutritional requirements, water demand and others (Ross, 2014) (Figure 7.1).

Livestock can also be observed easily and autonomously. Especially meaningful is the use of drones if the fields or grazing lands are very large or widely spread. Already in the 1980s remote midsize unmanned helicopters have been used for spraying applications in Japan (Yamaha Motor Australia, 2014). But even small UVAs are important. A new generation of small monitoring drones for protected plant cultivation inside greenhouses is currently in a development process (Van Hooijdonk, 2015). Many greenhouses nowadays have huge dimensions of 10 hectare or more. A permanent control of the plants is very laborious. Small drones will scan the crop regularly with cameras and specific optical sensors (Figure 7.2). Plant stress, pests and diseases, but also the development of the plants can easily be evaluated by growers or automated with the help of assessment software.

Figure 7.1. Unmanned aerial vehicle for remote sensing of field crops (A&L Canada Laboratories, London, Canada, 2015).

Figure 7.2. Autonomously flying quadcopter drone with sensors for growth and vitality of plants in greenhouses (Applied Drone Innovations Ltd, Rijswijk, the Netherlands, 2016).

7.3 Robots

One definition of robots is: 'A robot is a technical apparatus that usually takes mechanical work from humans. Robots can be both stationary and mobile machines. They are controlled by computer programs'. So robots shall primarily facilitate or even totally replace human work. A couple of robots already work in the field of agriculture. One of the most famous is probably the milking robot for cows and other dairy livestock. The animal decides itself if it is time to get milked. It enters a special cabin and the milking process starts. All the tasks a farmer did before are done autonomously by the milking robot. That means cleaning the udder and teats, pre-milking, milking and disinfecting. To make sure the cow stays patient, a feed dispenser is placed in front of it. But there's more to it than that. Every animal wears a special radio chip. By entering the robot the chip is scanned. During the milking process the milk is analysed immediately. The farmer gets information about the daily milk performance and even general health status, which gives him a base to act on (BouMatic Robotics, 2015). Chips and also video surveillance with integrated image and behaviour recognition make herd management much easier. But also mobile robots are in use. The self-driving feeding robot cares for physical well-being of cows by providing well mixed feed on demand. Another one herds cattle on the field and even leads them within the area (Dairy Science Group, 2013).

Besides livestock, horticulture is another application range for robots. The consumer price for fresh fruits and vegetables did not increase significantly during the past decade, but labour costs did. Nearly 50% of production costs for fruits and 35% for vegetables are for hired labour. Intensive horticultural crops require much more skilled labour compared to broad scale agriculture (Hewett, 2013). But often there is a problem of availability. Harvest time and yield is usually affected by weather and can vary from year to year, which makes it more complicated to arrange seasonal labour. Robots are patient. They work day and night with the same output quality. Actually in the horticulture sector a lot of robots are in use already and they are present in all stages of the value chain (Hewett, 2013). Applications include sowing, grafting, planting, potting and moving plants. During the cultivation period robots are used for irrigation, fertilising, spraying, weeding, cutting, mowing, moving and harvesting. But also at the stage of processing machines are responsible for sorting, cooling, packing, labelling, distribution and tracing.

Particularly for laborious, heavy and insanitary jobs like weeding, moving containers or spraying robots are a good solution and already available on the market. Some examples are the ecoRobotix weeding robot, the HV-100 for various distribution tasks or the Micothon EX spraying robot. But the trend is towards fully automated production units, especially in protected cultivation (New Growing System S.L., 2015). Large cultivation tables, fitted with technique for plant growth are moving independently through the greenhouse. Almost every process is predetermined, based on static or sensor controlled algorithms. No humans are involved in the entire physical workflow.

But not only robots are developed to manage plants. Also plants are adapted for better handling by robots. Big breeding companies focus on better machinable varieties. Some aspects are robustness against physical impacts, easier release of fruits or improved leaf positions for better harvest (phyllotaxy).

7.4 Vertical farming

One of the incontrovertible scenarios for mankind's future is that the world population will increase. At the same time cities will grow and rural zones will depopulate (United Nations, 2014). So the demand for fresh agricultural products will increase and it will be centred in the metropoles. The growing areas for the products will enlarge because of higher demands and they will move away from consumers into areas where less people will live. Furthermore crop production will take place on less fertile soils in the future, because settlements were mostly built on the most fertile soils, which were sealed more and more the bigger a city grows. 'Local', 'urban' and 'sustainable' are the major keywords in relation to prospective food production. The term vertical farming came up in the beginning of the 21st century, based upon the ideas firstly published by Dickson Despommier about stacked levels of production areas within a city (Despommier, 2010). Since then, plenty of projects and concepts were developed, but few were realised and still fewer work profitably.

The benefits of urban agriculture are clear: Fresh grown fruits, vegetables, herbs and even medical plants at the place of consumption without long transportation. But the challenges and problems are huge.

In the open field or in the greenhouse, sun provides light for free. In Central Europe we get approximately 700 watt of light energy per square meter at noon on a sunny summer day. We also get this energy in cities, but only at the highest unshaded level of a building. Rooftop farming was one answer here. This is one of the most popular and successful approaches. But even if all roofs of a city will be equipped, it would not be enough to feed all the dwellers and it would be very laborious and economically unviable to accomplish this decentralised production. So plants need to be cultivated in stacks, fully automated at one place and with the help of artificial light. But to realise this we need at least half of the sunlight's power at every single production level. So on one hand the main issue to solve is the high demand for electric energy. But plants actually process less than 1% of the provided energy. The rest changes to common heat. A vertical farm not only produces food, but also a lot of thermal energy. So, on the other hand, it is necessary to handle the big amounts of heat meaningful.

The energy supply of the future will consist of different conventional and renewable sources. To handle the various supplies efficiently so called smart grids must be installed as part of a smart city. In this way producers and consumers of electrical and thermal energies are in the position to operate within an energy network. Then demands and supplies are known and can be applied respectively. For vertical farms with their high demand for electrical energy it would be meaningful to 'switch on the lights' when the supply is high and the price is low. But, what to do with the emerging heat. The best scenario is to use it within the production of the farm for processing tasks, e.g. drying. Another idea is to implement aquacultures like fish, algae, pawn, sea shells and sea snails. They have a high potential to complete the range of products and some of them need warm water. In addition, water is a good buffer storage and transport medium for heat energy. Other thermophile organisms for food production are mushrooms and insects. Both only need heat and moisture, no light. If thermal energy is still left, it is possible to integrate it into the smart grid again and sell the heat to surrounding buildings and facilities like swimming baths and hospitals in winter time. During the summer heat can be stored underground.

Not only energy has to be used most efficiently, but also limited resources. Plant production needs huge quantities of water and fertilisers. Animal production is not possible without feed. To connect both by building closed circuits helps to reduce the essential inputs. Nutrient-rich wastewater from aquaculture can be upcycled by bacteria to reuse it as nutrient solution for plants. Plant refuse again is good feed for insects and some fishes. Of cause 'nothing comes from nothing' is valid in a vertical farm like anywhere else, but by creating almost closed cycles of materials and energies it is possible to produce food efficiently, sustainably and profitably.

7.5 Biotechnology

The definition of biotechnology by the Organisation for Economic Co-operation and Development is: 'Biotechnology is defined as the application of science and technology to living organisms as well as parts, products and models thereof, to alter living or non-living materials for the production of knowledge, goods and services' (OECD, 2001). In other words, the capabilities of biotechnology are not restricted to a single field, but are very multifarious. It is a cross-sectional technology that includes biology, biological chemistry, physics, chemistry, process engineering, material sciences and informatics. Biotechnologists not only do research on small and big organisms, plants, animals and humans, but on smallest parts like cells and molecules (BIOCOM AG, 2015). Common applications of biotechnology are used for producing beer, wine, bread, vinegar, cheese or yoghurt. And we have done it at least for 8,000 years (LaMar, 2006). Modern biotechnology can be divided into three different sections. The so-called RED Biotechnology deals with the development of new therapeutic and diagnostic medicinal methods. Here the human genome research is the base for new treatments. The WHITE Biotechnology is mainly connected with industry. Cosmetics, washing agents, chemistries, drugs and others are in the focus of this specialisation. Finally the GREEN Biotechnology handles agriculture, plants and plant related topics.

The genetic modification of organisms is in the centre of the green biotechnology. This mainly includes plants, but also animals and microorganisms. To conquer current and future challenges, e.g. climate change and population growth, secure and high yields are necessary. The genetic adaption of plants could be one way to realise this. Major aims of the green biotechnology are better resistance against drought, flood and coldness, stronger defence against pests and diseases and higher yields or simply higher health value.

There are different levels of manipulating plants. The most traditional is the common breeding. Here different genotypes of the same plant species are crossed, by dusting blossoms with pollen. The result is a countless variety of different new mixed genotypes. Choosing the best of them and cross them again and again could bring the expected feature. This way could be very long, especially if the young plants need a long time to bloom. The duration of breeding a new apple variety is 15 years and more. A faster method is to implant gens from other plant species by using genetic engineering. This is useful to transfer a specific attribute without modifying the rest of the genome. It is even possible to transfer gens from animals, microorganisms or viruses into plants.

In some cultures, especially in Western Europe, genetic engineering and genetically modified organisms are met with refusal. Here the so-called SMART breeding (Selection with Markers and Advanced Reproductive Technologies) or precision breeding has been developed. Smart breeding basically works in a similar way to traditional breeding. However, unlike traditional methods, in smart breeding the gene or gene variant responsible for a specific trait can be accurately identified using molecular biological procedures (DNA sequencing, PCR). It is then possible to test the offspring of a cross for the presence of the crossed gene, even before the actual trait is signalled by a changed external appearance. Only those plants which contain the desired gene are then grown on. The purpose of this is to introduce into crop plants genes from, e.g. wild populations which confer characteristics of interest to breeders (Schlegel, 2009).

Further areas of application of biotechnology are the production of substances with the help of microorganisms. Especially yeasts, bacteria, fungi and microalgae are adapted and used to produce specific substances for industry, medicine, agriculture or the food sector. One example is microalgae for producing bio fuels (Boldt and Graf, 2015).

7.6 Consumer perceptions

New technologies usually cause different reactions. One group of people basically like new developments and accept them quite fast. Another group is rather reluctant and rejects innovations. A small group is undecided or neutral. Differences in the acceptance of innovations also seem to be culturally conditioned. More than two thirds of the US-Americans have trust in the safety of their food. They appreciate biotechnology as an important tool for sustainable food production. (IFIC, 2014). In Northern Europe consumers are more critical towards biotechnology, especially genetically modified organisms are not accepted. Negative associations predominate possible benefits (e.g. improved taste, functional benefits, environmental benefits) (Grunert *et al.*, 2001). This attitude also applies to swaths of whole Europe.

Further differences in the acceptance of innovations can be related to the age and the educational level of people. It is known that the intelligence of vegetarians and vegans in the Western world is higher than the average (Gale *et al.*, 2007). But does it mean that the food pattern could be considered as a benchmark for intelligence? Generally, the knowledge about the food sector or technologies within the food sector is rather inadequate. This also applies to matters of common knowledge, like milking cows or where fruits come from. But this could also be a chance for the acceptance of new technologies in the future. If people like printing things in 3D, they maybe also like printing food, or like the idea of food factories in their direct neighbourhood.

References

BIOCOM AG, 2015. Was ist Biotechnologie? Available at: http://tinyurl.com/y777p52a.

Boldt, B. and Graf, P., 2015. Algentechnikum: Perfektes Licht für grüne Kerosin-Fabriken. Available at: https://tinyurl.com/y7ftnrfb.

BouMatic Robotics B.V., 2015. Milking robot MR-S1. Available at: https://tinyurl.com/yb5px5c5.

Condon, E.U. and Windsor, H.H., 1940. Driverless tractor plants crops in spirals. Popular Mechanics 74: 7.

Dairy Science Group, 2013. A robot amongst the herd. University of Sydney, Sidney, Australia.

Despommier, D., 2010. Vertical farm: feeding the world in the 21st century. St. Martins Press, New York, NY, USA, 311 pp.

Gale, C.R., Deary, I.J., Schoon, I. and Batty, G.D., 2007. IQ in childhood and vegetarianism in adulthood. British Medical Journal 334: 245-248.

Grunert, K.G., Lähteenmäki, L., Nielsen, N.A., Poulsen, J.B., Ueland, O. and Åström, A., 2001. Consumer perceptions of food products involving genetic modification: results from a qualitative study in four Nordic countries. Food Quality and Preference 8: 527-542.

Hewett, E.W., 2013. Automation, mechanisation and robotics in horticulture. Presentation at Workshop 'Emerging Postharvest Technologies'. Institute of Food, Nutrition and Human Health, Massey University, Albany, Auckland, New Zealand.

International Food Information Council (IFIC), 2014. Consumer perceptions of food technology survey. IFIC, Washington, DC, USA, 53 pp. https://tinyurl.com/y7xmv77l.

LaMar, J., 2006. Wine history. Science and social impact through time. Available at: https://tinyurl.com/cvwob7.

Mader, S., 2010. Center pivot irrigation revolutionizes agriculture. The Fence Post Magazine. Available at: https://tinyurl.com/uzmc3bx.

Murray, C.J., 2008. Deere takes next step toward driverless tractor. DesignNews. Available at: http://ubm.io/1pH9XE7.

New Growing System S.L., 2015. BabyLeaf MaxPro NGsystem. Available at: https://youtu.be/sEbb1dtBK0E.

Organisation for Economic Co-operation and Development (OECD), 2001. Biotechnology, single definition. Available at: https://tinyurl.com/y84pehxr.

Ross, P.E., 2014. Chris Anderson's expanding drone empire. Available at: https://tinyurl.com/nu5bc7k.

Schlegel, R.H.J., 2009. Dictionary of plant breeding. CRC Press, Boca Raton, FL, USA, 332 pp.

Tice, B.P., 1991. Unmanned aerial vehicles. The force multiplier of the 1990s. Airpower Journal 5: 18-25.

United Nations, 2014. World urbanization prospects. The 2014 revision: highlights. United Nations, New York, NY, USA, 28 pp.

Van Hooijdonk, R., 2015. Looking into the future. Presentation at GreenTech Summit 2015. Amsterdam, the Netherlands. Available at: https://youtu.be/7guXQRxHXgI.

Williams, M., 2002. Farm tractors. Features models from the world's leading manufacturers including John Deere, IH, Ford, Case, Mercedes-Benz, Massey-Ferguson. Lyons Press, Enderby, UK, 173 pp.

Yamaha Motor Australia, 2014. Yamaha RMAX Type IG/Type II unmanned helicopter. Available at: https://tinyurl.com/yc6687h7.

8. Modernisation of traditional food processes and products

L. Lipan, L. Sánchez-Rodríguez, M. Cano-Lamadrid, J. Collado-González, L. Noguera-Artiaga, E. Sendra and A.A. Carbonell-Barrachina*

Departamento Tecnología Agroalimentaria, Escuela Politécnica Superior de Orihuela, Universidad Miguel Hernández de Elche, Carretera de Beniel, km 3.2, 03312 Orihuela, Alicante, Spain; angel.carbonell@umh.es

Abstract

Tradition and innovation are generally considered antonyms; this antagonistic relationship makes innovation of traditional foods very challenging. However, it is possible to innovate in traditional foods leading to what we call 'tradfoods'. The culture, tradition, sensory quality, and health benefits can be maintained, but modernizing the manufacturing process leads to safer products, and the new tradfoods are adapted to needs and demands of current consumers. If these basic requirements are fulfilled, consumers are willing to accept the new tradfoods (with innovations in packaging, environmental sustainability, etc.) and they are even willing to pay more for them.

Keywords: fat reduction, food safety, innovation, internationalization, salt reduction

8.1 Introduction

A traditional food can be defined as a product made in a non-industrial, traditional and/ or unique environment that is characterised by a production in small batches and with a low degree of mechanisation (Kupiec and Revell, 1998). This definition implies that the traditional product presents a wide variety but, at the same time, a solid identity (Rason *et al.*, 2007). Because of this characteristic diversity and personality, the consumer can easily distinguish among traditional and industrial foods (Fancois *et al.*, 2000). Traditional foods are a significant element in European culture, identity, and heritage; they contribute to the development and sustainability of rural areas and increase the supply of products of high organoleptic quality (Guerrero *et al.*, 2009). The authors of this chapter have decided to use the expression 'tradfoods' for 'new' traditional foods, which ensure food safety, modernization of their production chains, but at the same time assure that their organoleptic quality is the same; this last condition is essential.

There is a clear relationship among the availability of raw materials and traditional foods. For instance, most Greek rural families prepared fermented meat products from pork meat and fat, chopped and mixed together with salt and seasonings, shortly before Christmas (when they slaughtered their home-grown pigs, Panagou *et al.*, 2013). In a similar way, Spanish families prepared *turrón* at Christmas time at their homes because almonds and honey are fully available that time of the year. Traditional products are subjected to a self-consumption economy and, thus, rely on the maximum use of available resources; they are also a method of conservation for raw materials, which are mainly produced at a specific time of the year (seasonality of production), whereas food needs exist throughout the whole period (Jordana, 2000).

Originally, traditional foods were basically consumed at home; however, now traditional products are sold in gourmet shops, and traditional gastronomy is mostly found in restaurants (Jordana, 2000).

In a context of change of the consumption culture, the interest of consumers in traditional products with high quality or 'own personality' has increased in the European Union and other developed countries. It has been recently shown that European consumers believe that the hard work and expensive production of traditional products are justified by their specific sensory attributes and health characteristics (Almli *et al.*, 2011), and want to buy and consume them. The consumers' perception of the quality of traditional foods with 'own personality' greatly depends on their personal preferences, level of experience, cultural influences (Iaccarino *et al.*, 2006), demographic and psychological characteristics (Oude Ophuis and Van Trijp, 1995), product authenticity perceived by the consumer, and the expectation of high quality. This last factor can be also affected by the geographical origin (Stefani *et al.*, 2006), price, nutritional information (Kähkönen and Tuorila, 1998), traditional processing technology (Rason *et al.*, 2007) and, probably, by cultural conditions, such as the irrigation conditions under which the plant products are obtained (especially for countries or regions where water is a scarce resource).

Traditional products, while distributed and consumed at small levels, are highly appreciated by consumers, who are willing to pay more for a product in which they clearly recognise special characteristics, well worth the extra money. Thus, most traditional foods have high prices and their commercialisation is mainly limited to gourmet or specialised channels (Jordana, 2000). Problems arise when traditional foods want to be promoted and sold in a larger scale, which could provide an extra income to their producers, because not all consumers are aware of the uniqueness of the products and their willingness to pay is not so high.

8.2 Is innovation of traditional foods possible?

As previously mentioned, the commercialisation of traditional foods is drastically limited to local markets and gourmet shops. However, there are good perspectives for the growth of these products if some challenges are addressed (Jordana, 2000): (1) communication, a traditional product is exotic in any other market and has to be advertised; (2) legal protection is not guaranteed in the international markets; (3) quality assurance, this should be a priority because it is included in the definition of traditional foods; and (4) innovation, which is required to ensure safety and adjustment to the new nutritional requirements of consumers, among other factors.

In this chapter, we will focus on the challenges of innovation and sanitary quality. In the past, the small size of family companies producing traditional foods and their limited technical education has sometimes resulted in microbiological problems. This has drastic consequences for the viability of the company and the future of traditional foods; food safety is a must and should not be jeopardised at all. If qualitative insecurity is associated with a particular brand, consumers will stop purchasing it (Jordana, 2000), and it will seriously damage the products of that producer, region or even country. When food safety alerts happen, the products of a particular country are banned at the international level, and traditional foods are not an exemption.

Tradition and innovation are considered antonyms, which makes innovation of traditional foods very challenging. In this way, it may seem a paradox talking about innovative tradfoods, but it is possible, if, as mentioned before, some requirements are accomplished. The culture, tradition, and health benefits of this type of products can be maintained, but at the same time, it is possible to improve especially their nutritional characteristics to adapt to consumers' needs and demands. In the past, traditional foods have been basically developed in rural environments where a great degree of physical work was performed and then a high caloric intake was required. However, this is not the case for urban life; therefore, the nutritional characteristics must be adjusted to the current needs of consumers, but without jeopardising the essence of the product, its quality. This is essential for the health of our society, because there is no doubt that the number of consumers, even young ones (school students) suffering obesity is drastically increasing (Zahnd *et al.*, 2015).

In this way, if consumers demand health and/or safety innovations in the traditional food sector, and traditional producers need to increase their sales, it is understandable that they undertake these innovations ensuring their own survival, by producing new traditional products more in tune with market requirements (Jordana, 2000). In general, both the agro-food sector and consumers are open to innovations in traditional food products, and agree that the common prerequisite is to preserve the traditional character of the food (Kühne *et al.*, 2010). However, innovation in the tradfoods sector is limited because of the size and low technology of the companies; the agri-business sector, and particularly the tradfood sector, is characterised by a large number of micro-, small-, and medium-sized-enterprises (SMEs). In general, tradfoods can be characterised by factors, such as: (1) their key production steps are performed in a certain area at national, regional or even local level; (2) they are authentic in their recipe (mix of ingredients), origin of raw material, and/or production process; (3) they are commercially available for at least 50 years; and (4) they are part of the gastronomic heritage. Unfortunately, the majority of these products do not have any form of origin or quality label (Kühne *et al.*, 2010) that guarantees that raw materials, preparation processes, and/or final products are controlled.

The main fields of innovations in tradfoods are packaging innovations and changes in product composition, product size, and new ways of using the product (Gellynck and Kühne, 2008). However, process innovations are less common, given their impact on the authentic identity of the product and its production process. Feasible applications relate to improving the production process to guarantee quality and traceability. However, innovations to make the process faster and continuous are also possible; for example, *turrón* companies in Spain are transforming the manufacture of *Alicante turrón* into a continuous process. Finally, the implementation of market and organizational innovations can be valuable for tradfoods, but their potential is not yet realised or recognised by all chain members in this sector (Gellynck and Kühne, 2008).

Recent efforts to assess the pan-European consumers' interpretation of the concept of tradfoods through qualitative research in six European countries (Belgium, Italy, Norway, Poland, France, and Spain) resulted in the following definition.

> A traditional food product is a product frequently consumed or associated with specific celebrations and/or seasons, normally transmitted from one generation to another, made accurately in a specific way according to the gastronomic heritage, with little or no processing/manipulation, distinguished and known because of its sensory properties and associated to a certain local area, region or country.
>
> (Guerrero *et al.*, 2009)

Consumers were particularly positive towards packaging innovations. A different package does not modify the core characteristics of these products, but can provide the sought benefits, e.g. longer shelf life (Guerrero *et al.*, 2009). Further, approval of nutrition-related innovations stemmed from a positive link with pronounced tangible benefits and corresponded with earlier studies on specific tradfoods (Caporale and Monteleone, 2004; Cayot, 2007). Next, positive acceptance with regard to convenience-oriented innovations was associated with opportunities, if this did not involve too remarkable changes in the product. Finally, product innovations with implications for the sensory properties were strongly rejected (Cayot, 2007), because this is the core of the high quality of tradfoods (Figure 8.1).

In agreement with all the above, it has been reported that improving the quality of tradfoods through better selection of ingredients and/or raw materials, better uniformity of the product, and improved packaging is indicated as primary innovation activity in the traditional food chain (Kühne *et al.*, 2010). These same authors reported that consumers showed interest in innovations regarding food safety and potential health benefits; this was especially true for reducing fat in foods, especially dairy products. Additionally, it has been shown that individual portions are well accepted because of the tendency towards a higher share of single living individuals in society. However, innovations towards pre-cooked food or ready-to-eat dishes do not meet consumers' demands, because traditional foods are consumed at special occasions, in a social context with the family and/or friends (Kühne *et al.*, 2010). One of the more remarkable findings of this study was that tradfoods should be supported by a label that guarantees their origin, quality, specificity, and authenticity (Kühne *et al.*, 2010). Finally, it was pointed out that females and urban consumers are more prone to accept innovations in tradfoods (Guerrero *et al.*, 2009; Kühne *et al.*, 2010).

As a summary of this section, it can be stated that innovation in tradfoods may be seen in terms of improvements in safety, healthiness, and/or convenience attributes of the products

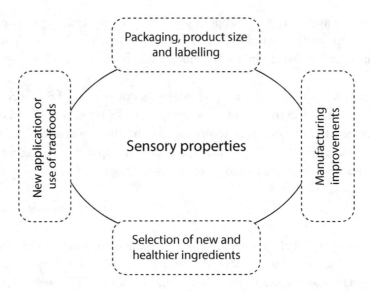

Figure 8.1. Schematic abstract showing the different strategies to innovate in the sector of the traditional foods, showing that the core (the most important and non-touchable part) of their quality is the sensory quality.

(Guerrero *et al.*, 2009). They mainly pertain to packaging innovations and changes in product composition, product size and form or new ways of using the product (Gellynck and Kühne, 2008). Other innovations may relate to product image through packaging, labelling, brands, and/or additional certifications in accordance with the indications of disciplinary production regulations.

8.3 Types of innovation of traditional foods

In the rest of this chapter, different ways of innovation in tradfoods will be presented and discussed according to the existing scientific literature. The innovation options will include:
- transformation of non-continuous into continuous processing;
- establishing better control strategies over growing or processing conditions;
- reduction of costs;
- adjustment of nutritional properties, and functionalization;
- ensuring food safety.

Besides listing and discussing the different types of innovation to be implemented, examples or case studies will also be provided. In this way, it is expected that the readers can identify tradfoods from their home countries in which similar opportunities exist and could be implemented. Because the authors are from Spain and have cooperated with many companies preparing tradfoods, most of the examples will be from Spanish products.

8.3.1 Transformation of non-continuous into continuous processing

One of the important transformations needed in the tradfood sector is converting non-continuous processing into a continuous system, with appropriate control over the working conditions. As previously mentioned, this change should never jeopardise the organoleptic quality of the product, but for sure will improve the homogeneity of the product within the same batch and from batch to batch.

A case study of this type of innovation is *turrón de Alicante* (Figure 8.2e), which is a Spanish tradfood prepared by heating up honey, sugar, and water until reaching a caramel type product (Figure 8.2a). After this, ovalbumin is added as a whitening agent, and later toasted almonds are added (Figure 8.2b) and manually mixed (Figure 8.2c) due to the high viscosity of the product. Then, the product is moulded (Figure 8.2d) and finally packed (Vázquez-Araújo *et al.* 2006, 2007, 2008b, 2012;). The composition of this Spanish Christmas confection (Figure 8.2e) is optimized by using '*Marcona*' almonds and rosemary or orange and lemon blossoms honeys (Vázquez-Araújo *et al.*, 2008a,b, 2009; Verdú *et al.*, 2010). However, their manufacturing is a non-continuous system with production in batches in the heating containers. The bigger companies with big customers are trying to work on the improvement of their manufacturing process. The first steps of this manufacturing process (production of caramel and mixing of the almonds) are well developed and do not negatively affect the sensory quality of the product. However, the automatic preparation of the product in a thin layer is still causing problems with the texture, because hot almonds are fragile and some of them break during this step. But in a very near future this problem will be solved and the product will be available in the tradfoods markets with exactly the same quality than non-continuous *turrón*, but with reduced costs and avoiding a risky handling (charges of the small heating containers, their cleaning, and the hand-manipulation of the 'hot' product by the workers).

On the other hand, there is a key unit operation called '*boixet*' in *Jijona turrón* in which an inversion phase happens and the texture of the product clearly depends on it (Pérez-Munuera *et al.*, 2001). This is the limiting step in transforming this system into a continuous one, because it lasts for about 1-1.5 h at a temperature between 50-80 °C. There is no possible innovation in this limiting unit operation, in which a phase inversion occurs (a water-in-

Figure 8.2. Example of a non-continuous manufacturing process of a Spanish traditional food, *turrón de Alicante*, which can be transformed into a continuous system with detailed control over the working conditions.

oil system is transformed into an oil-in-water system), because it will drastically reduce the quality of the final product (Martínez-Navarrete *et al.,* 1996; Martinez and Chiralt, 1996; Martínez *et al.,* 1997).

As a summary, it can be stated that there are options to innovate in the manufacturing of tradfoods, but there are key unit operations which cannot be modified because the organoleptic quality of the product will be negatively affected.

8.3.2 Establishing better control over growing and producing conditions

The working hypothesis in this subsection is that if a better control is applied during growing of the raw materials or in the different unit operations during production of tradfoods, the quality of the final product will be improved.

One example of the first case is ensuring the use of traditional cultivars as raw ingredients or the growing of the called 'hydroSOStainable or hydroSOS' vegetables. These products are cultivated under controlled deficit irrigation, limiting irrigation water during non-sensitive growing periods of the plants. These products behave as 'traditional foods' because they have a solid identity (higher content in bioactive compounds, higher intensity of some sensory attributes, etc.), and besides, they are environmentally friendly because they optimize the use of a very valuable resource in the world, and especially in the Mediterranean region, the water (Cano-Lamadrid *et al.,* 2015; Carbonell-Barrachina *et al.,* 2015). This is an example that an appropriate control over the growing conditions can lead to better quality products. Later hydroSOS almonds can be used for the manufacturing of tradfoods, such as *turrón* in Spain or *torrone* in Italy, improving their quality and respect to the environment.

Following with the tradfoods included in the previous section, if the toasting temperature of almonds is controlled and specific temperature programs are developed for specific almond cultivars, the amounts and composition of the pyrazines developed during this step will be the most appropriate ones and will lead to a higher consumer acceptance (Vázquez-Araújo *et al.* 2008a, 2009).

8.3.3 Reduction of costs

Innovations in the processing of tradfoods should lead to reduced costs and enhanced homogeneous quality (Jordana, 2000). The idea is to go from a totally artisanal elaboration to the implementation of effective and already proven technologies, which do not entail modifications in the intrinsic characteristics of the product. Their task is to ensure an increased productivity, a reduced price, but without compromising the quality and the essence.

The cost reduction can be reached through different factors: (1) the use of 'only' local raw materials; (2) the use of most effective unit operations (e.g. continuous systems); and (3) longer shelf-life and lower spoilage of the produced foods by using new or emerging preservation technologies, which simultaneously improve the safety of the products.

8.3.4 Adjustment of nutritional properties and functionalisation

It has been reported that four out of every five new products reaching the market nowadays offer consumers some sort of advantage regarding health benefits: a high content of fibre, a low content of saturated fats, a high content of calcium, etc. (Jordana, 2000). It is imperative that companies elaborating tradfoods consider these tendencies, if they want to be successful in the very competitive sector of foods with extra-value. Perhaps it is just a matter of getting a better knowledge of the products and their composition, and highlighting their 'positive' aspects. A second option is to replace some of the raw materials with others to adjust the product to the new market requirements while keeping the product elaboration process and providing the product with unique new quality characteristics or directed to fit special needs (gluten-free products). This should be only done in those cases in which the authenticity is not jeopardised by the changes in the raw materials.

In this way 'functional' or 'improved' tradfoods can be designed (Vintilă, 2015):

▸ to reduce caloric intake to limit weight gain and reduce obesity risk;
▸ to restrict intake of added sugar, and especially of refined sugar;
▸ to avoid intake of saturated fats and cholesterol from foods of animal origin associated with cardiovascular disease;
▸ to reduce the intake of salt to improve blood pressure;
▸ to recommend or support consumption of fruits, vegetables, oils, and cereals rich in vitamins, minerals, and fibres, which is linked with positive health benefits, including reduced risk of well-spread 'modern' diseases (e.g. cardiovascular disease and cancer);
▸ to produce allergen-free products; and
▸ to include new 'healthy' ingredients.

Traditional foods with a high caloric content are often related to high contents of sugars and/or fats. For instance, *turrón* contains more than 50% of toasted almonds (product with ~ 50% oil content), and about 20-30% of sugars (Verdú *et al.*, 2007). Thus, *turrón* is an appropriate model to check how companies producing traditional foods are dealing with the reduction of the caloric intake and added sugars. The first recommendation that these companies are giving to their consumers is that a balanced diet should be followed and that it is better to eat small portions of the product and enjoy them, than just eat, eat, and eat, and, then, suffer the health problems linked with its consumption. In this way, many companies are preparing individual portions (~20 g) instead of 250-300 g bars. This is a very clever strategy because it is promoting a healthy consumption and it is adjusting their products to the new reality of the society in which many people live alone. Another strategy is reducing the added sugars by more healthy compounds, such as inulin (Garcia-Garcia *et al.*, 2013). However, and as previously mentioned, reducing the content of almond is not an option, because this will negatively affect the taste and flavour of the product and consumers will reject it.

Lipids are an important part of the human diet; both saturated and unsaturated fats have important functions and should be part of a healthy diet, but in appropriate amounts. There is a clear relationship among high intake of saturated fats with accumulation of cholesterol, metabolic syndrome, and obesity (Downs *et al.*, 2015; Nilsen *et al.*, 2015; Ríos-Hoyo *et al.*, 2015). A further technological reduction of fat amounts can be achieved by means of the addition of starch and hydrocolloids with notable water absorption (Nasrin and Anal, 2014; Tahergorabi *et al.*, 2015; Viebke *et al.*, 2014); however, the appearance of the final products is not always the desired one.

The impact of diet on healthiness is often highlighted by researchers and the World Health Organization (WHO). In particular, charcuterie products are viewed negatively from both medical and nutritional viewpoints because of the high percentage of some ingredient such as salt, fat, and spices or additives, such as nitrite. In this way, there have been links established between processed meat and colorectal cancer (WHO, 2015). There are different options to replace nitrates and nitrites in meat products; our research team is working on the use of different extracts of natural products, such as leaf-vegetables, herbs, and pomegranate, with success; but still more research is needed.

The sodium (Na) and sodium chloride (NaCl) intake is related to hypertension and cardiovascular diseases. Salt is well known for its ability to preserve the product from the degradative action of spoilage bacteria; it can contribute to the palatability of the final product and it carries out the solubilisation of salt-soluble proteins. All these reasons make the reduction the amount of salt difficult. A crucial aspect trying to replace the salt is the sensory acceptability of the product. In general, sensory parameters such as brightness, greasiness, graininess, colour, and compactness are dependent on the added salt. The research on this is a hot topic, and many researchers are focussed on reducing the content of salt in many traditional foods, such as cheese (Kamleh *et al.,* 2015), meat products, such as *jamón serrano* (Mora-Gallego *et al.,* 2016; Wu *et al.,* 2014; Ying *et al.,* 2016), and fish products, such as Cantabrian anchovies (Jittrepotch *et al.,* 2015; Sasaki *et al.,* 2014), among other products.

The Mediterranean diet pyramid recommends the consumption of table olives on a daily basis due to nutritional benefits associated with this fruit (Bach-Faig *et al.,* 2011). Traditional elaboration processes of table olives involving fermentation include a step of immersion in a saline solution, where fermentation takes place. One approach to reduce NaCl in foods may be the utilization of salt substitutes (other chlorides) alone or combined. Combinations of sodium, potassium, calcium and magnesium salts have been used to perform olive fermentation in the presence of reduced levels of sodium chloride during the fermentation and in the final product (Bautista-Gallego *et al.,* 2013a,b,c). Therefore, this is an active field of research and traditional table olives 'low in sodium' of the highest sensory quality will be available in a near future.

Allergens are naturally occurring proteins responsible for causing food allergy in consumers which presence should be clearly indicated on food labels, even when only trace amounts may be present. Ensuring the absence of several allergens, mainly gluten, lactose, nuts, milk and egg ingredients offer an added value for allergic consumers. In this sense, traditional producers with high safety standards will be able to benefit from such extra-value (Jiménez-Saiz *et al.,* 2015; Schaub and Vercelli, 2015).

The micronutrients, such as minerals (Ca, Mg, K, and Na), are essential for the metabolism and tissue functions of humans. To accomplish the daily recommended doses, it is necessary to have a balanced and varied diet. Dairy products are one of the important sources of Ca, but in some poor countries the consumption of these products is limited due to the costs and lactose intolerance. However, a potential contributor to the micronutrients intake is the consumption of fish products, e.g. anchovies (*Engraulis ringens*), sardines (*Sardina pilchardus*), and mackerel (*Scomber scombrus*), that are particularly rich in Ca, vitamin A, Fe, and Zn (Albrecht-Ruiz and Salas-Maldonado, 2015). The consumption of this small fish provide a cheap source of micronutrients and protein in the diet (Eurofish, 2012), and enhance micronutrient intakes in vulnerable populations, such as children, pregnant and lactating women (Kabahenda *et al.*, 2011, 2014).

Traditionally, 'Asian dried' anchovies (*ikan bilis*) have been consumed intact as a whole fish, including the bones; this way of eating anchovies is very healthy because of the nutritional benefits of the whole fish being available for the consumer (Henry *et al.*, 2015). Meanwhile, the modern method of consuming anchovies is to eat them in a cleaned, eviscerated form, after the removal of head, skeleton and entrails. This way of eating anchovies drastically reduces the intake of Ca, Fe, and Zn (Figure 8.3), although it has other advantages (high organoleptic quality, reduced Na content, and increased K content). The problem is that the modernization of the traditional method can have unintentional nutritional consequences. A minor change in the dietary habits of eating cleaned anchovies may lead to a reduction in micronutrients intake. This reinforces the need for caution when we modernize our traditional eating habits (Henry *et al.*, 2015). It is important to mention that this type of anchovies are different from those consumed in Europe, for example Cantabrian anchovies, in which the whole fish is matured by the fish digestive enzymes while completely covered by salt; later, the fish is cleaned, and immersed normally in olive oil.

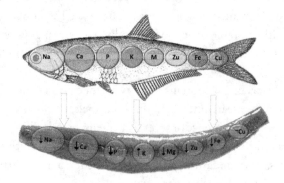

Figure 8.3. Effect of modernisation of food habits on the mineral contents of fish.

Traditional plants could be used as a good source of functional foods, including tradfoods. Many plants, which have been traditionally used only as medicines, are now being incorporated into traditional foods to improve their health benefits (Rivera *et al.,* 2010). These authors describe the potential use of white sapote (*Casimoroa edulis*), jicama (*Pachyrhizus spp.*), amaranth (*Amaranthus hypochondriacus*), sweet fennel (*Foeniculum vulgare*), oregano (*Lippia graveolens*), pitahaya (*Hylocereus* sp.), agave (*Agave Americana*), pelitre (*Heliopsis longipes*), and purslane (*Portulaca oleracea* L.).

In some cases, new healthy ingredients can be used to complete the 'concept' of existing tradfoods. The most popular types of *turrón* (Alicante and Jijona) contain toasted almonds, 'honey', and sugar, as the main components. *A la piedra turrón* consists only of toasted almonds, sugar, and very small amounts of lemon scrapes, and cinnamon (Narbona *et al.,* 2010); thus, it does not contain honey, and the idea behind its functionalization was to replace this ingredient by other bee-products (propolis or royal jelly). In this way, propolis (0.05%) and royal jelly (1.0%) were added to *a la piedra turrón* (García-García *et al.,* 2012; Narbona *et al.,* 2010). The functionalized *a la piedra* turrón is healthier than the traditional version because of the health benefits associated with consumption of propolis or royal jelly, but the sensory quality is not jeopardised because these ingredients were added below their detection thresholds, and consumers will not recognise their flavour. Finally, this same research group studied the effects of replacing sugar by inulin on the blood serum lipids, and demonstrated that *turrón de Jijona* with inulin decreased the LDL cholesterol level in volunteers consuming this type of confection during 5 weeks (Garcia-Garcia *et al.,* 2013).

The addition of functional ingredients into cheese, such as probiotic bacteria, may be accepted only if tangible benefits are delivered to consumers (Braghieri *et al.,* 2014). For example, the incorporation of probiotic strains (*Bifidobacterium longum* and *Bifidobacterium lactis* or *Lactobacillus acidophilus*) into the cheese matrix of *Scamorza* cheese (Albenzio *et al.,* 2013a,b). The matured *Scamorza* cheeses contained significantly higher contents of some amino acids compared to control cheese; the amino acids included glutamine, serine, arginine, isoleucine, and leucine, which are useful in improving gut function, mucosal integrity and vascular development. On the other hand, a negative effect of omega-3 inclusion in traditional cheese in Spain was found on product liking (Hersleth *et al.,* 2005a,b).

8.3.5 Ensuring food safety

It is essential that enterprises, producers' associations, and local, national, and international authorities control and guarantee the safety of tradfoods. When consumers buy a food, including tradfoods, they 'should not care' about food safety, because it is something that should be taken for granted. In any case, it is also true that consumer concern about food safety increases world-wide, and they are much more interested in not only 'what food to eat' but also 'what food not to eat'. This worry is based on the fact that pathogens borne by contaminated foods sicken millions, and kill thousands, each year worldwide. Thus, it is the responsibility of all the stakeholders involved in the production and distribution of tradfoods to assure food safety. Otherwise and if a health alert or crisis comes out and it is associated with tradfoods, the future of all these items will be seriously jeopardised.

Due to the massive emigration, traditional foods from developing countries are required all over the world. However, to be produced and exported they need to address the following challenges: (1) cover the needs of local people; and (2) produce higher amounts of the products to be exported.

North African countries have a rich tradition in food technology, and many traditional foods of animal and plant origin are widely consumed and highly appreciated. They are still mainly prepared at a household level under 'poor sanitary conditions' and marketed through informal routes, with no control over food safety. Consequently, their consumption is anticipated to put the public health at risk (Benkerroum, 2013). This is a clear example of the urgent need to draw to the attention of stakeholders (legislators in North African countries) into the health risks associated with their consumption, and consequently, to take the necessary measures to reduce such risks. At the same time and after improving the safety of these traditional foods, there could be a nice market opportunity for them in other countries or regions, such as the European Union due to the high number of African people living in these countries and willing to buy and consume their tradfoods. The consumption of tradfoods from their home-countries strengthens connections with their origin places, and helps in keeping their way of living and their future development (Settanni and Moschetti, 2014).

A recent review was conducted about traditional fermented foods of North African countries, and discussed the safety of products produced in countries such as Mauritania, Morocco, Algeria, Tunisia, Libya, Egypt, and Sudan (Benkerroum, 2013). It has been reported that the main associated pathogens with traditional 'dairy products' are *Staphylococcus aureus*, *Salmonella* spp., *Escherichia coli*, *E. coli* O157, *Botulinum*, *Bacillus cereus*, *Listeria monocytogenes*, *Campylobacter* spp., and *Clostridium* spp.; with traditional 'meat products' are: *S. aureus*, *Clostridium botulinum*, *Clostridium perfringens*, *E. coli* O157:H7, *Campylobacter* spp., etc.; and with traditional 'plant products' are: *C. botulinum*, *C. perfringens*, *L. monocytogenes*, *B. cereus*, and *Pseudomonas enteritis*.

As a summary of the situation of North African countries, which is a perfect example for other developing countries or regions worldwide, the unsafe conditions of tradfoods are mainly due to one or more of the following reasons (Benkerroum, 2013):

▶ Lack or inappropriate veterinary care in farms; weak prophylactic program.
▶ Slaughtering, carcass dressing, evisceration, and meat cutting are done under poor sanitary conditions.
▶ The production of traditional meat products is done in small farms, butcheries, or food shops, where the sanitary conditions are not appropriate, and none of the GMP (good manufacturing practice), GHP (good hygiene practice), or HACCP (hazard analysis and critical control point) are implemented.
▶ Inadequate conditioning, packaging, and storage conditions.

It is important to finish this section indicating that not all tradfoods from Northern African countries are risky and that they are a fantastic proof of the gastronomic richness of this region. However, there are many things to do before these products can be fully safe and widely distributed at international scale.

Recurring incidents with pesticide residues in vegetable-based tradfoods and regular breaches in food hygiene have placed food safety high on the policy agenda in Southeast Asia, including Vietnam (Reardon *et al.,* 2007). In the rapidly developing economies of Southeast Asia, food safety concerns resulting from industrialization, urbanization and increasing distance among producers and consumers need special attention and are of concern.

On the other hand, there are tradfoods which are considered very healthy and even functional, for instance *kimchi*. In the last decades, attention has focused on traditional Korean foods, which are based on slow food fermentation. Generations ago, such foods were established as safe and healthy foods. *Kimchi* was presented in the Health Magazine in 2006 as one of the world's five healthiest foods. However, improvements are needed because its manufacturing is moving from small-scale to mass production (Lee *et al.,* 2012).

Greek tradfoods have evolved over centuries according to local culture and artisanal practices in small family-owned facilities and constitute a major part of the so-called Mediterranean diet. Greek tradfoods are a model for other Mediterranean foods, such as Italian and Spanish ones. Consumer' concern for food safety and high demand for traditional foods is becoming an important challenge for the industry (Panagou *et al.,* 2013). According to these authors, the conditions have drastically changed over the last years, where the production has shifted from an artisanal practice to an industrial level under strict processing and hygienic conditions. Now, it can be considered that Greek traditional foods have a good safety record, but a key question then arises, can these foods still be considered as traditional foods? Has anybody checked the sensory quality of these products? Can consumers distinguish among the old traditional and the new-industrial traditional foods? (Panagou *et al.*, 2013). Again it is essential that these questions are answered by those producing the new tradfoods at the very beginning and just before starting their commercialisation; if this is not the case and products are fully safe but their quality is not high, consumers will buy them once, but they will not repeat because of the high price which should be directly linked to high quality.

8.4 Conclusions

The authors want to conclude this chapter by showing a successful example of how innovation can be implemented. For a protected designation of origin (PDO) of cheese, the bond with its own traditional characteristics is important because protected foods must show the link with 'authentic and unvarying local methods' and 'tradition'. This indication implies that such products also have distinctive sensory characteristics, which are connected to the traditional production methods, and guarantees the consumer a sensory quality which lasts over time. However, Parmigiano-Reggiano cheese is a clear example of how a traditional cheese has experienced changes in the 20th century due to accelerated rate of modernization, but still holds its high sensory quality. Its quality is related to its granular texture, intense characteristic taste, and simultaneous sweet but rather fatty flavour (Zannoni, 2010). Consequently, it can be concluded that modernization of the traditional foods can be implemented leading to tradfoods of the utmost sensory quality, but ensuring food safety and large scale of production; however, research and resources need to be invested to reach this goal. If these changes are properly designed and implemented, consumers are more than ready to accept these 'improved' tradfoods. In this way, the opinion of Italian consumers on different types of innovations affecting traditional Italian cheeses was checked (Pilone *et al.*, 2015). Consumer's willingness to pay for new attributes (e.g. sustainability and shelf life extension) was high, and synergies across food policies oriented to innovation, product quality, and environmental sustainability, were created and will result in a better future for the cheese industry.

References

Albenzio, M., Santillo, A., Caroprese, M., Braghieri, A., Sevi, A. and Napolitano, F., 2013a. Composition and sensory profiling of probiotic Scamorza ewe milk cheese. Journal of Dairy Science 96: 2792-2800.

Albenzio, M., Santillo, A., Caroprese, M., Ruggieri, D., Napolitano, F. and Sevi, A., 2013b. Physicochemical properties of Scamorza ewe milk cheese manufactured with different probiotic cultures. Journal of Dairy Science 96: 2781-2791.

Albrecht-Ruiz, M. and Salas-Maldonado, A., 2015. Chemical composition of light and dark muscle of Peruvian anchovy (*Engraulis ringens*) and its seasonal variation. Journal of Aquatic Food Product Technology 24: 191-196.

Almli, V.L., Næs, T., Enderli, G., Sulmont-Rossé, C., Issanchou, S. and Hersleth, M., 2011. Consumers' acceptance of innovations in traditional cheese. A comparative study in France and Norway. Appetite 57: 110-120.

Bach-Faig, A., Berry, E.M., Lairon, D., Reguant, J., Trichopoulou, A., Dernini, S., Medina, F.X., Battino, M., Belahsen, R., Miranda, G., Serra-Majem, L. and Mediterranean Diet Foundation Expert Group, 2011. Mediterranean diet pyramid today. Science and cultural updates. Public Health Nutrition 14: 2274-2284.

Bautista-Gallego, J., Arroyo-López, F.N., Romero-Gil, V., Rodríguez-Gómez, F., García-García, P. and Garrido-Fernández, A., 2013a. Microbial stability and quality of seasoned cracked green aloreña table olives packed in diverse chloride salt mixtures. Journal of Food Protection 76: 1923-1932.

Bautista-Gallego, J., Moreno-Baquero, J.M., Garrido-Fernández, A. and López-López, A., 2013b. Development of a novel Zn fortified table olive product. LWT – Food Science and Technology 50: 264-271.

Bautista-Gallego, J., Rantsiou, K., Garrido-Fernández, A., Cocolin, L. and Arroyo-López, F.N., 2013c. Salt reduction in vegetable fermentation: reality or desire? Journal of Food Science 78: R1095-R1100.

Benkerroum, N., 2013. Traditional fermented foods of North African countries: technology and food safety challenges with regard to microbiological risks. Comprehensive Reviews in Food Science and Food Safety 12: 54-89.

Braghieri, A., Girolami, A., Riviezzi, A.M., Piazzolla, N. and Napolitano, F., 2014. Liking of traditional cheese and consumer willingness to pay. Italian Journal of Animal Science 13: 155-162.

Cano-Lamadrid, M., Girón, I.F., Pleite, R., Burló, F., Corell, M., Moriana, A. and Carbonell-Barrachina, A.A., 2015. Quality attributes of table olives as affected by regulated deficit irrigation. LWT – Food Science and Technology 62: 19-26.

Caporale, G. and Monteleone, E., 2004. Influence of information about manufacturing process on beer acceptability. Food Quality and Preference 15: 271-278.

Carbonell-Barrachina, A.A., Memmi, H., Noguera-Artiaga, L., Del Carmen Gijón-López, M., Ciapa, R. and Pérez-López, D., 2015. Quality attributes of pistachio nuts as affected by rootstock and deficit irrigation. Journal of the Science of Food and Agriculture 95: 2866-2873.

Cayot, N., 2007. Sensory quality of traditional foods. Food Chemistry 102: 445-453.

Downs, S.M., Marie Thow, A., Ghosh-Jerath, S. and Leeder, S.R., 2015. Aligning food-processing policies to promote healthier fat consumption in India. Health Promotion International 30: 595-605.

Eurofish, 2012. Overview of the world's anchovy sector and trade possibilities for Georgian anchovy products. Eurofish, Copenhagen, Denmark.

Fancois, M., Sylvander, B. and Hossenlop, J., 2000. Typicité et mode de production: une typicité fermière? In the socioeconomics of origin labelled products in agri-food supply chains: spatial, institutional and coordination aspects. INRA editions, Le Mans, France.

Garcia-Garcia, E., Narbona, E., Carbonell-Barrachina, A.A., Sanchez-Soriano, J. and Roche, E., 2013. The effect of consumption of inulin-enriched Turrón upon blood serum lipids over a 5-week period. International Journal of Food Science and Technology 48: 405-411.

García-García, E., Narbona, E., Jurado, P., Burló, F., Roche, E. and Carbonell-Barrachina, Á.A., 2012. Volatile composition of 'A la piedra' turrón enriched with royal jelly. Italian Journal of Food Science 24: 132-139.

Gellynck, X. and Kühne, B., 2008. Innovation and collaboration in traditional food chain networks. Journal on Chain and Network Science 8: 121-129.

Guerrero, L., Guàrdia, M.D., Xicola, J., Verbeke, W., Vanhonacker, F., Zakowska-Biemans, S., Sajdakowska, M., Sulmont-Rossé, C., Issanchou, S., Contel, M., Scalvedi, M.L., Granli, B.S. and Hersleth, M., 2009. Consumer-driven definition of traditional food products and innovation in traditional foods. A qualitative cross-cultural study. Appetite 52: 345-354.

Henry, C.J., Bi, X., Lim, J. and Lau, E., 2015. Mineral decline due to modernization of food habits. Food Chemistry 190: 194-196.

Hersleth, M., Ilseng, M.A., Martens, M. and Næs, T., 2005a. Perception of cheese: a comparison of quality scoring, descriptive analysis and consumer responses. Journal of Food Quality 28: 333-349.

Hersleth, M., Ueland, Ø., Allain, H. and Næs, T., 2005b. Consumer acceptance of cheese, influence of different testing conditions. Food Quality and Preference 16: 103-110.

Iaccarino, T., Di Monaco, R., Mincione, A., Cavella, S. and Masi, P., 2006. Influence of information on origin and technology on the consumer response: The case of soppressata salami. Food Quality and Preference 17: 76-84.

Jiménez-Saiz, R., Benedé, S., Molina, E. and López-Expósito, I., 2015. Effect of processing technologies on the allergenicity of food products. Critical Reviews in Food Science and Nutrition 55: 1902-1917.

Jittrepotch, N., Rojsuntornkitti, K. and Kongbangkerd, T., 2015. Physico-chemical and sensory properties of Plaa-som, a Thai fermented fish product prepared by using low sodium chloride substitutes. International Food Research Journal 22: 721-730.

Jordana, J., 2000. Traditional foods: challenges facing the European food industry. Food Research International 33: 147-152.

Kabahenda, M., Mullis, R.M., Erhardt, J.G., Northrop-Clewes, C. and Nickols, S.Y., 2011. Nutrition education to improve dietary intake and micronutrient nutriture among children in less-resourced areas: a randomized controlled intervention in Kabarole district, western Uganda. South African Journal of Clinical Nutrition 24: 83-88.

Kabahenda, M.K., Andress, E.L., Nickols, S.Y., Kabonesa, C. and Mullis, R.M., 2014. Promoting dietary diversity to improve child growth in less-resourced rural settings in Uganda. Journal of Human Nutrition and Dietetics 27: 143-151.

Kähkönen, P. and Tuorila, H., 1998. Effect of reduced-fat information on expected and actual hedonic and sensory ratings of sausage. Appetite 30: 13-23.

Kamleh, R., Olabi, A., Toufeili, I., Daroub, H., Younis, T. and Ajib, R., 2015. The effect of partial substitution of NaCl with KCl on the physicochemical, microbiological and sensory properties of Akkawi cheese. Journal of the Science of Food and Agriculture 95: 1940-1948.

Kühne, B., Vanhonacker, F., Gellynck, X. and Verbeke, W., 2010. Innovation in traditional food products in Europe: do sector innovation activities match consumers' acceptance? Food Quality and Preference 21: 629-638.

Kupiec, B. and Revell, B., 1998. Speciality and artisanal cheeses today: the product and the consumer. British Food Journal 10: 236-243.

Lee, G.I., Lee, H.M. and Lee, C.H., 2012. Food safety issues in industrialization of traditional Korean foods. Food Control 24: 1-5.

Martínez-Navarrete, N., Fito, P. and Chiralt, A., 1996. Influence of conditions of manufacture and storage time on the textural characteristics of Xixona turrón. Food Control 7: 317-324.

Martinez, N. and Chiralt, A., 1996. Glass transition and texture in a typical Spanish confectionery product: Xixona turron. Journal of Texture Studies 26: 653-663.

Martínez, N., Chiralt, A. and Fito, P., 1997. Transport phenomena in the phase inversion operation of 'Xixona turrón' manufacture. Journal of Food Engineering 32: 313-324.

Mora-Gallego, H., Guàrdia, M.D., Serra, X., Gou, P. and Arnau, J., 2016. Sensory characterisation and consumer acceptability of potassium chloride and sunflower oil addition in small-caliber non-acid fermented sausages with a reduced content of sodium chloride and fat. Meat Science 112: 9-15.

Narbona, E., García-García, E., Vázquez-Aráujo, L. and Carbonell-Barrachina, A.A., 2010. Volatile composition of functional 'a la piedra' turrón with propolis. International Journal of Food Science and Technology 45: 569-577.

Nasrin, T.A.A. and Anal, A.K., 2014. Resistant starch: properties, preparations and applications in functional foods, functional foods and dietary supplements: processing effects and health benefits. Wiley Blackwell, Hoboken, NJ, USA, pp. 227-253.

Nilsen, R., Høstmark, A.T., Haug, A. and Skeie, S., 2015. Effect of a high intake of cheese on cholesterol and metabolic syndrome: results of a randomized trial. Food and Nutrition Research 59: 27651.

Oude Ophuis, P.A.M. and Van Trijp, H.C.M., 1995. Perceived quality: a market driven and consumer oriented approach. Food Quality and Preference 6: 177-183.

Panagou, E.Z., Nychas, G.J.E. and Sofos, J.N., 2013. Types of traditional Greek foods and their safety. Food Control 29: 32-41.

Pérez-Munuera, I., Lewis, D.F. and Lluch, M.A., 2001. Microstructure changes during the Xixona turrón manufacture analyzed by light microscopy. Food Science and Technology International 7: 473-478.

Pilone, V., De Lucia, C., Del Nobile, M.A. and Contò, F., 2015. Policy developments of consumer's acceptance of traditional products innovation: the case of environmental sustainability and shelf life extension of a PGI Italian cheese. Trends in Food Science and Technology 41: 83-94.

Rason, J., Martin, J.F., Dufour, E. and Lebecque, A., 2007. Diversity of the sensory characteristics of traditional dry sausages from the centre of France: relation with regional manufacturing practice. Food Quality and Preference 18: 517-530.

Reardon, T., Henson, S. and Berdegué, J., 2007. 'Proactive fast-tracking' diffusion of supermarkets in developing countries: implications for market institutions and trade. Journal of Economic Geography 7: 399-431.

Ríos-Hoyo, A., Cortés, M.J., Ríos-Ontiveros, H., Meaney, E., Ceballos, G. and Gutiérrez-Salmeán, G., 2015. Obesity, metabolic syndrome, and dietary therapeutical approaches with a special focus on nutraceuticals (polyphenols): a mini-review. International Journal for Vitamin and Nutrition Research 84: 113-123.

Rivera, G., Bocanegra-García, V. and Monge, A., 2010. Traditional plants as source of functional foods: a review. CYTA – Journal of Food 8: 159-167.

Sasaki, T., Araki, R., Michihata, T., Kozawa, M., Tokuda, K., Koyanagi, T. and Enomoto, T., 2014. Removal of cadmium from fish sauce using chelate resin. Food Chemistry 173: 375-381.

Schaub, B. and Vercelli, D., 2015. Environmental protection from allergic diseases: from humans to mice and back. Current Opinion in Immunology 36: 88-93.

Settanni, L. and Moschetti, G., 2014. New trends in technology and identity of traditional dairy and fermented meat production processes: preservation of typicality and hygiene. Trends in Food Science and Technology 37: 51-58.

Stefani, G., Romano, D. and Cavicchi, A., 2006. Consumer expectations, liking and willingness to pay for specialty foods: do sensory characteristics tell the whole story? Food Quality and Preference 17: 53-62.

Tahergorabi, R., Matak, K.E. and Jaczynski, J., 2015. Fish protein isolate: development of functional foods with nutraceutical ingredients. Journal of Functional Foods 18: 746-756.

Vázquez-Araújo, L., Chambers, D. and Carbonell-Barrachina, Á.A., 2012. Development of a sensory lexicon and application by an industry trade panel for turrón, a European protected product. Journal of Sensory Studies 27: 26-36.

Vázquez-Araújo, L., Enguix, L., Verdú, A., García-García, E. and Carbonell-Barrachina, A.A., 2008a. Investigation of aromatic compounds in toasted almonds used for the manufacture of turrón. European Food Research and Technology 227: 243-254.

Vázquez-Araújo, L., Verdú, A., Enguix, L. and Carbonell-Barrachina, A.A., 2008b. Investigation of aromatic compounds in Alicante and Jijona turrón. European Food Research and Technology 227: 1139-1147.

Vázquez-Araújo, L., Verdú, A., Murcia, R., Burló, F. and Carbonell-Barrachina, A.A., 2006. Instrumental texture of a typical spanish confectionery product Xixona turron as affected by commercial category and manufacturing company. Journal of Texture Studies 37: 63-79.

Vázquez-Araújo, L., Verdú, A., Navarro, P., Martínez-Sánchez, F. and Carbonell-Barrachina, A.A., 2009. Changes in volatile compounds and sensory quality during toasting of Spanish almonds. International Journal of Food Science and Technology 44: 2225-2233.

Vázquez-Araújo, L., Verdú, A., Miquel, A., Burló, F. and Carbonell-Barrachina, A.A., 2007. Changes in physico-chemical properties, hydroxymethylfurfural and volatile compounds during concentration of honey and sugars in Alicante and Jijona turrón. European Food Research and Technology 225: 757-767.

Verdú, A., Vázquez-Araujo, L. and Carbonell-Barrachina, Á.A., 2007. Mathematical quantification of almond content in Jijona turrón. European Food Research and Technology 226: 301-306.

Verdú, A., Vázquez-Araujo, L., Miquel, A., Martínez-Sánchez, F. and Carbonell-Barrachina, A.A., 2010. Discriminant analysis of almond cultivars used in turrón. Italian Journal of Food Science 22: 76-82.

Viebke, C., Al-Assaf, S. and Phillips, G.O., 2014. Food hydrocolloids and health claims. Bioactive Carbohydrates and Dietary Fibre 4: 101-114.

Vintilă, I., 2015. Actual state and perspectives of Christian religious dietary laws and certification in Romania. Trends in Food Science and Technology 45(1): 147-152.

World Health Organisation (WHO), 2015. IARC Monographs evaluate consumption of red meat and processed meat. IARC, Lyon, France.

Wu, H., Zhuang, H., Zhang, Y., Tang, J., Yu, X., Long, M., Wang, J. and Zhang, J., 2014. Influence of partial replacement of NaCl with KCl on profiles of volatile compounds in dry-cured bacon during processing. Food Chemistry 172: 391-399.

Ying, W., Ya-Ting, J., Jin-Xuan, C., Yin-Ji, C., Yang-Ying, S., Xiao-Qun, Z., Dao-Dong, P., Chang-Rong, O. and Ning, G., 2016. Study on lipolysis-oxidation and volatile flavour compounds of dry-cured goose with different curing salt content during production. Food Chemistry 190: 33-40.

Zahnd, W.E., Rogers, V., Smith, T., Ryherd, S.J., Botchway, A. and Steward, D.E., 2015. Gender-specific relationships between socioeconomic disadvantage and obesity in elementary school students. Preventive Medicine 81: 138-141.

Zannoni, M., 2010. Evolution of the sensory characteristics of Parmigiano-Reggiano cheese to the present day. Food Quality and Preference 21: 901-905.

9. Emerging technologies in food processing

V. Glanz-Chanos[1], S. Shayanfar[2]* and K. Aganovic[3]

[1]Aarhus University, Bartholins Alle 14, 8000 Aarhus C, Denmark; [2]General Mills Inc., 9000 Plymouth Ave N, Minneapolis, MN 55427, USA; [3]Deutsches Institut für Lebensmitteltechnik, Prof.-von-Klitzing-Str. 7, 49610 Quakenbrück, Germany; shima.shayanfar@genmills.com

Abstract

In this chapter, a number of emerging technologies will be described that are being employed, or are about to be employed, in food processing. The technologies described in this chapter are the ones that have the potential to result in new food products with benefits for consumers, especially benefits that are related to the trends presented elsewhere in this book, including the trends towards healthy, authentic, sustainably produced food products that are convenient to use. Therefore, by necessity the following review cannot be comprehensive. The technologies have been grouped according to the main benefits achieved by them, knowing that some technologies can result in multiple benefits. The chapter draws heavily on work carried out in the EU project RECAPT[1], and additional input from the EU project HighTech Europe[2].

Keywords: emerging, alternative, non-thermal, advanced, food processing

---

[1]   Retailer and Consumer Acceptance of Novel Technologies and Collaborative Innovation Management; see www.recapt.org.
[2]   First European Network of Excellence in Food Processing; see www.foodtech-portal.eu.

Klaus G. Grunert (ed.) **Consumer trends and new product opportunities in the food sector**
DOI 10.3920/978-90-8686-852-0_9, © Wageningen Academic Publishers 2017

9.1 Introduction

Food processing is when a raw agricultural commodity is being subjected to different steps (e.g. washing, cleaning, milling, cutting, chopping, heating, pasteurizing, blanching, cooking, canning, freezing, drying, dehydrating, fermenting, mixing, packaging, or other procedures) that alter the food from its natural state. Processing also may include the addition of other ingredients to the food, such as preservatives, flavours, nutrients, and other food additives or substances approved for use in food products, such as salt, sugars, and fats (IFIC, 2014). The technologies covered in this chapter are grouped in the following way:

▸ Technologies improving food safety by thermal or non-thermal processes (e.g. use of high pressure, electro-magnetic fields such as pulsed electric fields, ohmic heating, cold plasma, electron beam irradiation and infrared heating).

▸ Texturizing and forming methods that obviate or mitigate the use of any other additives such as enzymes or emulsifiers (e.g. hydrodynamic pressure technology, high pressure homogenization) and advanced cutting methods such as ultrasonic cutting.

▸ Gentle/mild processing methods (e.g. super critical fluid extraction) that improve product quality and can add value in terms of a 'clean label' product.

▸ New packaging solutions with a potential positive effect on sustainability by reducing food waste (e.g. biodegradable packaging materials, intelligent and active packaging, radio frequency identification and edible coatings).

9.2 Improving food safety by thermal or non-thermal processes

9.2.1 High pressure processing

High pressure processing (HPP; also sometimes high hydrostatic pressure (HHP)), is a mild preservation method for inactivating the harmful bacteria and spoilage microorganisms. In this method the already packaged product is subjected to high hydrostatic pressure (up to 6,000 bars) in a high pressure vessel for few minutes (mostly 1-5 min). Successful examples of application of HPP include ham slices, snacks, fish, ready to-eat meats (cold cuts), fresh juice, prepared fruits and vegetables, picked crabmeat and oysters, fruit smoothies, guacamole, chicken strips, and salsa. The main benefit compared to conventional technologies is that the product is not exposed to heat and usually no additional chemical is used, resulting in a more natural product with enhanced taste and flavour, less preservatives (such as salt, etc.) (Aganovic *et al.*, 2014). Further HPP is a post-packaging preservation step, with reduced recontamination risk. However, as the HPP is a batch process, higher processing costs are involved. However, despite the high processing costs HPP is a batch process. The other shortcomings include possible colour/texture change in raw meat products and requirements of specific packaging materials.

Pressure application can be used for the preservation of solid and liquid food products at ambient temperature, provided they have sufficient water and no air voids. The process takes only few minutes and meets the highest hygienic requirements, as the products are treated in their final package. The shelf life of different products can be extended from few days to up to few weeks. Furthermore, making use of pressure-induced structure modification allows developing food products with new textural properties. For wheat starch, for example, a swelling of the granules is observed at pressure levels of 300 MPa.

In protein-based products, depending on the pressure and protein, different phenomena can occur. High pressure of 5,000-6,000 bars leads to protein denaturation and solidification. Pressures in the range of 3,000 bars often lead to protein solubilisation and hydration. Lower pressure levels of 1000 bars allow protein conditioning and gelation. This effect is normally achieved in presence of salt. Thus the technology has potential for development of salt reduced meat products. The technology can also be used to replace or assist 'cold' cooking of meat products. Contrary to conventionally cooked products, pressure-treated food exhibit reduced cooking loss, thus providing for a high product yield.

9.2.2 Pulsed electric field technology

Pulsed electric field technology (PEF) is another alternative method for processing food by means of brief pulses of a high electric field. For processing, a substance is placed between two electrodes, then the PEF is applied. The treatment causes structural rearrangements, changes in conductivity and increased cell membrane permeability to ions and macromolecules. As such increase in permeability is, presumably, related to the formation of pores on the cell membrane, the phenomenon is often called 'electroporation' or 'electropermeabilisation' (Ivorra and Rubinsky, 2010). Examples of applications are pasteurized liquid and semi-liquid, pumpable products such as juice, smoothies, dairy products, soups, and cold brewed coffee. It can also be used as processing aid for mitigating other processing methods due to electroporation effect in products such as tomato peeling, potato cutting, freezing and extraction of valuable compounds. The main benefits are fresh like products attributes, enhanced organoleptic quality, no chemical additives, inactivation of vegetative micro-organisms including yeasts, spoilage micro-organisms and pathogens in liquid products such as juices, milk and soups, juice yield improvement, and enhanced extraction of functional components from different tissues. PEF shortcomings include cold storage requirements and aseptic filling necessity.

The cell membranes of microorganisms, plant or animal tissue can be made permeable by using a PEF. This effect can be used for a variety of purposes in food and bio-processing. The application of this technology is energy-efficient, environmentally friendly and commercially viable. The process has been successfully commercialized. The treatment capacity of the continuous, industry-ready equipment can be up to 10,000 l/h for liquid media preservation and 50 t/h for cell disintegration. Additionally, microbial decontamination of liquids can be achieved bellow conventional pasteurization temperatures. The natural freshness and appearance of the product as well as the vitamin content are better retained. Due to a targeted effect on cell membranes, the functional and technological properties are not affected. Furthermore, the method can be used for the pre-treatment of fruits, vegetables or oil seeds, algae or cell cultures for the production of functional ingredients. Compared to time-consuming and costly enzymatic maceration, this method increases the yield and also retains pigments, antioxidants and vitamins.

9.2.3 Cold plasma technology

Cold plasma is another non-thermal technology for surface decontamination of plastic, glass bottles, trays, caps and cans before filling. The principle of atmospheric cold plasma is based on a pulsed radio frequency discharge in gases to which an electric charge has been applied (nitrogen, argon, air) at atmospheric pressure creating no equilibrium plasma on the treated surface. The benefits are rapid, effective, energy-efficient decontamination with reduced damage to the product. As for shortcomings, bacteria in deeper biofilm layers survive better after the plasma treatment than without or at small layer of biofilm, the rate of inactivation, e.g. of *Salmonella* Typhimurium, is inversely proportional to initial bacterial concentration. Cold plasma electrodes may degrade over time, metals that are evaporated at the electrode surface may end up as unintended additives.

Cold plasma gas can be used to decontaminate surfaces without significantly damaging the surface, but little is known about the critical parameters for this technology in the commercial setting. The equipment relies on application of gas discharge technology, and has traditionally mainly been used for surface batch sterilisation of medical equipment in hospitals. However, the potential for wide-scale application of food grade cold plasma gas is substantial, and the quality and shelf-life of cold plasma decontaminated foods and packaging materials is significantly better than for foods produced using traditional preservation technologies.

9.2.4 Electron beam irradiation

Electron beam irradiation, also known as electron beam (eBeam) processing, is a non-thermal treatment of food products and ingredients using eBeam accelerators for food preservation by breaking of DNA chain in spoilage and pathogenic organisms. The irradiation technology today has been approved in over 50 countries in Asia, the Americas, Africa and Europe;

and used commercially for a variety of foods including frozen and fresh meat, frozen and fresh poultry, grains, spices, shellfish, and food ingredients, phytosanitary-treated fruits and vegetables, and sterilized packaging materials in aseptic packaging systems. It is quick, reliable, compatible with most materials, and can be applied after packaging, producing fresh-like products with flavour and colour retention and no nutrients loss (Pillai and Shayanfar, 2015). Irradiation at doses higher than 10 kGy (which rarely happen in food processing) may cause undesirable changes in the organoleptic properties of food including off-flavour, discoloration or tissue softening (Sebranek *et al.*, 2014). Generally speaking, the capital cost for irradiation is slightly higher compared to other alternative methods. This lies on relatively high investment and start-up costs. On the other hand, irradiation is a low labour continuous technology, and the input requirement for conventional energy is low. Therefore, processing costs are lower than the other technologies. However, both retailers and consumers have been sceptical towards this technology. Irradiation is not accepted by the organic industry, although organic agriculture supports the use of treatments that least change the commodity from its untreated state. Taking into account that irradiation leaves no residue, causes minimal chemical or physical changes to the commodity and is tolerated by more fresh commodities than any other treatment, organic industry might eventually accept irradiation (Cannon *et al.*, 2012).

Preservation of products such as herbs and spices or other dry products that cannot be easily decontaminated can be achieved by eBeam. The method can be applied in batch or continuous operations to treat packed, palletized or bulk commodities. Depending on the dose, an extension of the shelf life, a reduction of pathogenic or spoilage microorganisms, a disinfestation or sprouting inhibition can be achieved. The low temperature applied in this technology ensures that the freshness and heat-sensitive nutrients in food products are preserved. It is an environmentally friendly and time effective alternative to traditional thermal decontamination technology.

Application of eBeam can inhibit microbial growth in food through the direct or indirect damage to the physiological metabolism and chemical reactions performed by microorganisms, leading to their injury or death. By applying eBeam the generated energy transfer within the microorganisms results in the destruction of the chemical and molecular bonds. The cells are therefore no longer able to perform their normal physiological metabolism activities. On the other hand, the eBeam may ionize the water molecules to produce unstable free radicals, which consequently damage other cellular metabolic pathways to promote intracellular oxidation, resulting in cell injury and death. The decontamination effect of the eBeam on food is commonly affected by factors such as the species of indicator microorganisms and the chemical composition of the food. High-dose irradiation treatment can also damage the sensory characteristics and nutritional quality of food. The decontamination mechanism of the eBeam must therefore be properly understood in order to achieve a better irradiating effect under appropriate conditions. A good irradiation plan based on the developed methodology would increase the shelf life and improve the quality of food products.

9.2.5 Ohmic heating

Ohmic heating (OH) is a continuous process for heating products with the purpose of tempering, thawing, pasteurization and sterilization. In OH food is heated by passing electrical current through it. Due to resistive properties of food, the electrical energy is dissipated into heat, which results in rapid and uniform heating. The benefits are fast, uniform, direct, volumetric heating of the entire mass of the homogeneous food material, which results in less overheating and greater quality. The technology is suitable for processing large particulate foods (up to 1 inch) that would be difficult to process using conventional heat exchangers. Less cleaning requirements are needed due to reduced product fouling on the food contact surface. It is environmentally friendly, with up to 90% conversion efficiency, heating rates of up to 50 °C/min. When heating liquids containing particles, it results in less fouling and burnt particles. Maintenance costs are reduced because of the lack of moving parts. However, there are some shortcomings considered for this technology too. The installation costs of OH is high. Furthermore, the presence of fat globules is non-conductive (due to lack of water and salt) and any pathogenic or spoilage bacteria that may be present in these globules may receive less heat treatment than the rest of the substance, as the electrical conductivity of a substance is important. Further, the product is in direct contact with electrodes, compared to for example microwave heating as another form of volumetric heating. Inhomogeneous foods are difficult to process and require adoption of the electrodes for parts with different conductivity/resistivity (e.g. ham with visible meat and fat parts).

9.3 Texturising/forming methods that mitigate use of additives

9.3.1 Hydrodynamic pressure technology

Hydrodynamic pressure technology (HDP), also known as Shockwave technology, is a novel hydrodynamic technique in which ultrasound high-energy shock waves in water cause instantaneous tenderization of a tissue, e.g. meat. Meat tenderness is an important quality parameter of meat and is achieved by using different methods, such as biological tenderization, electrical stimulation, hot-boning, aging, etc., but is difficult to achieve in e.g. cattle derived from milk production. The application of shockwaves is caused by dissipation of energy and mechanical stress at the boundary areas of materials with different sound velocity and acoustic impedance. The benefits are uniform tenderness of low quality meat parts, little negative impact on meat quality traits such as water-holding capacity, juiciness, flavour, and fresh meat colour, and improving the processing characteristics of moisture-enhanced meat products. A shortcoming is that the product needs packaging prior to processing, necessitating shockwave resistant packaging materials.

It is possible to use mechanical force for the tenderization of beef, pork and poultry meat. The application of underwater shockwaves has been shown to be a highly efficient physical method to effect meat tissue disintegration and accelerate meat maturation. Shockwaves with the required energy density can be generated by explosives, but also by underwater discharge of electrical energy. This electro-hydraulic generation of shockwaves allows for an energy efficient and safe application.

When using this technology, packaged meat is submerged in a vessel and exposed to shockwaves. The mechanical stress caused by these shockwaves as well as secondary, biochemical reactions accelerate maturation and reduces the curing time. The total energy input is in a range of a few kJ/kg, corresponding to less than 1 °C temperature increase.

Shockwave treated beef meat shows a decrease in cutting force after cooking. Whereas conventionally a maturation time of 14 days is required, shockwave application allows decreasing this period to 7 days. The product quality is increased and storage and distribution efforts are reduced.

9.3.2 Ultrasonic cutting

Ultrasound is cyclic sound pressure with a frequency greater than the upper limit of human hearing which can be used to induce mechanical vibrations and is produced by converting electrical energy into mechanical energy which is quite high in frequency. Conversion from electrical to mechanical energy is accomplished by applying electrical energy to a transducer. Generally, an ultrasonic cutting system consists of a generator, a transducer, an amplifier and a sonotrode (blade), which is capable of applying frequencies in the low ultrasonic frequency range of 20-100 kHz. The technique can be applied to soft and hard cheese, snack bars, health bars, ice cream, bakery, frozen cakes and pies, frozen fish, prepared meats, pizza, gum and candy, fresh and frozen vegetables and bones. The benefits are improved cutting quality and precision, reduced reaction forces in a continuous cutting operation, it avoids wetting of cut material (compared to water jet cutting), avoids noises and smoke and air contamination (compared to laser cutting), and avoids adhering food products to sonotrode, which makes cleaning more convenient and produces less product waste. Shortcomings are probable changes in the material's mechanical properties particularly in food with a high fat content. Also, the increased temperature may cause fluidization along the contact layer and there is a tendency to burn at the cut surface during ultrasonic cutting in case of poor temperature control at the blade.

Ultrasonic waves are an 'inaudible sound', the frequency of which generally exceeds 20 kHz. A 20-kHz frequency means that a certain medium vibrates 20,000 times per second. The ultrasonic cutter vibrates its blade with amplitude of 10-70 μm in the longitudinal direction. The vibration is microscopic, so it cannot be seen. Because of this movement, the ultrasonic cutter can easily cut different materials, including food products. This unique method of cutting saves time and results in finer cuts in all materials, no matter how soft or hard the structure is. Additionally, there will not be any adhesion of the food products on the surface of the blade, which obviates blade fouling and fastens the cutting process.

9.3.3 High pressure homogenisation

High pressure homogenization (HPH) is a purely mechanical process, which is evoked by forcing a liquid product through a narrow gap (the homogenizing nozzle) at very high pressure. The gap of the UHPH valve is significantly smaller (2-5 μm) compared to the valve gap of a traditional homogeniser (10-30 μm) (Floury *et al.*, 2004a). This allows higher maximum operating homogenisation pressures of up to 400 MPa, compared to pressure achieved using traditional homogenisers of up to 100-150 MPa. In the HPH the liquid flows under pressure for a very short time (1-10 s) to the narrow disruption valve and then expands along with instantaneous pressure drop at the exit of a homogenising valve, resulting in formation of very small fat droplets (Desrumaux and Marcand, 2002; Floury *et al.*, 2004b). Sudden pressure release results in a significant temperature increase caused by shearing and partial conversion of dynamic energy into heat in the range of 15-20 °C/100 MPa (for aqueous matrices) (Donsi *et al.*, 2009). Thus, the HPH technology arises as a potential technology for production of very fine emulsions, allowing for clean label and food products with reduced amount fat, emulsifiers, stabilisers and thickeners. Using the additional side effect of heating, the technology can be used at the same time for preservation. Using very fine emulsions it can facilitate the use of less fat without compromising creaminess, thus offering a healthier option. This feature is of value for low fat products such as mayonnaise, spreads and ice creams. Besides, the technology can be used in pharma industry, cosmetics as well as bio-based technologies for extraction of intracellular material, due to cell disruption under high shear forces. In food products, it can improve the rheological properties of ice cream mixtures and stabilize cream liquors. As it is only applicable to pumpable products, the viscosity of the raw material is important.

Dispersion methods are applied for retaining susceptible components (such as vitamins, flavour, etc.) in food matrices, for the formation of interfaces with specific properties or for the generation of multi-phase food structures. Novel dispersion principles based on microporous membranes as well as ultra-high pressure homogenizers have been developed to produce tailor-made dispersed systems. These tools open up new possibilities for the generation and stabilization of multi-phase systems. Encapsulation of susceptible or unpleasantly tasting substances or the formation of emulsions and foams with a narrow size distribution are some examples of many potential applications. Homogenization is a state-of-the-art continuous process used in the dairy industry for emulsification and many other applications. Pressure-induced unfolding of large molecules, e.g. hydrocolloids can be used for the modification of their functional and technological properties.

9.4 Gentle/mild processing methods

9.4.1 Super critical fluid extraction

That a soluble material is extracted from a food material by a solvent is a well-developed process (e.g. sucrose from beet or cane; coffee from coffee beans). The significant difference in super critical fluid extraction is that the solvent is a supercritical fluid, that is, a substance at a temperature and pressure above its critical point where distinct gas and liquid phases do not exist. An example of a supercritical fluid in food use is carbon dioxide or water. Possible applications include cocoa butter from cocoa beans; decaffeination of coffee; cholesterol removal from oils/fats; recovery of alkaloids. The benefits are short time processing, low flammability, low toxicity, benign environmental effects, no solvent waste that requires remediation, no yeast needed for baking, and the possibility of processing heat sensitive foods. However, it is a quite new technology and involves high capital cost.

The physical-chemical properties of fluids can be varied with increases in pressure and temperature. Near the critical point, a decrease of the dielectric constant results in an increasing solvent power for non-polar substances. For instance, supercritical carbon dioxide is used to obtain hop extracts or for the decaffeination of coffee. Depending on the temperature and pressure – for water the critical point is at 374 °C and 21 MPa – its ion product is increased and hydrolytic reactions are catalysed. Exposure to supercritical fluids enables the extraction of valuable substances as well as the hydrolysis of biopolymers such as cellulose, starch or proteins.

By-products such as peels, pulps or protein-containing solutions resulting from food processing are often used for animal feed. To improve their sustainability and commercial viability, it is desirable that they are used for the generation of energy or within the food chain, but this calls for suitable processing methods.

Supercritical fluid hydrolysis can be used to convert biopolymers into functional food ingredients or to allow for enzymatic or microbial fermentation. Due to its specific physical-chemical properties, supercritical water is an optimal reaction media for biopolymer hydrolysis. In a temperature range between 150 and 250 °C and at a pressure between 10 and 20 MPa, cellulose or proteins are hydrolysed within seconds. Short holding times allow the processing in small, continuous reactors. No chemicals are required, thus eliminating any disposal or neutralization issues. Heat recovery supports an energy efficient processing approach.

9.4.2 Infrared heating

Infrared (IR) is part of the electromagnetic spectrum in the wavelength range between 0.5 and 1000 mm. IR light falls to spectra between visible light and microwaves, and can be divided to near- (0.78-1.4 μm), mid- (1.4-3.0 μm) and far-IR (3.0-1000 μm). The food components absorb energy effectively in the far-IR region 3-1000 mm), resulting in heating of food systems.

IR has several uses in processing of solid foods, such as drying, thawing, roasting, blanching, baking, cooking or preservation. Current uses include browning of meat (sandwich meat, hamburgers, Doner kebabs, etc.), baking of bread and confectionery; heating of some desserts (e.g. burnt sugar surface of crème brûlée), filled chocolates, etc. The benefits are a direct and fast heat transfer, no heating of the surrounding air, better retention of nutrients, reduced processing time and energy required, uniform temperature on the product surface, high heat fluxes, decreased flavour loss,. However, it is mainly efficient for solid products and has low penetration power.

The preservation of dry bulk products, surfaces or tools often is difficult because of heat-transfer limitations. For such matrices, the energy transfer by irradiation can be a highly efficient solution. With the selection of a suitable type of irradiation and the respective wavelength (IR, ultraviolet or ionizing), targeted effects can be achieved due to the different inactivation mechanisms and penetration depths. IR or ultraviolet light can be used for a microbial decontamination. High intensity IR sources allow hyper-thermization of surfaces. Exposure to short-time, high-intensity IR irradiation results in a microbial inactivation while retaining the product quality. Possible application examples include surfaces of meat, sausages or carcasses as well as bread.

9.5 Advanced packaging methods

9.5.1 Edible coatings

Edible coatings are transparent food based materials that cover the food item and act as barrier to humidity and oxygen. The idea of using edible coatings has been obtained from skin of fruits and vegetables that can act as an edible barrier against the foreign contaminants. These are thin layers of edible materials which restrict loss of water, oxygen and other soluble material of food.

Moreover, these films can be used as a host for additives in the conservation of the properties of the product or simply in order to improve its appearance. They can be applied to meat products, seafood, ready meals, fruits and vegetables, snacks, dry products. Considering their natural origin, they reduce environmental pollution by mitigating the application of polymeric packaging materials. These coatings can reduce physical, colour and texture changes in fresh cuts. Further, they can protect against contamination, reduce gas diffusion, control oxidation and respiration reactions, as well as serve as a carrier for food additives like antioxidants or antimicrobials, thus reduce the decay without affecting quality of the food. However, they are still costly to use.

Today, use of edible coatings is a common issue that is beneficial to protect nutrients material of food especially fruits and vegetables and provide a long durability. They may develop nutritional value and present bactericidal effects. These films can be placed on the surface of food products through different methods such as dipping, spraying and fluidized bed systems.

9.5.2 Active packaging

Active packaging materials are materials and articles that are intended to extend the shelf-life or to maintain or improve the condition of a packaged food. Active packaging is accurately defined as 'packaging in which subsidiary constituents have been deliberately included in or on either the packaging material or the package headspace to enhance the performance of the package system'. This phrase emphasizes the importance of deliberately including a substance with the intention of enhancing the food product. Active packaging is an extension of the protection function of a package and is commonly used to protect against oxygen and moisture.

Active packaging actively interacts in a positive way, changing the condition of the packaged food to this end:

▸ extend shelf life;
▸ improve food safety;
▸ improve sensory properties, while maintaining quality of food.

Active packaging technologies include some physical, chemical, or biological action which changes interactions between a package, product, and/or headspace of the package in order to get a desired outcome. Depending on the active packaging mechanism, they can be divided in three major groups: (1) absorber (e.g. oxygen absorber); (2) emitters (release certain compounds into packaging, CO_2, antioxidants, etc.); and (3) special systems (self-heating, modifiers for microwave application, etc.). Based on the physical shape, they can be integrated into packaging in form of sachets, labels, films, trays, caps or closures, and be even integrated in films as an active barrier.

The range of application covers variety of foods, such as meat, seafood, fruits and vegetables, pasta, cheese, bakery goods, ready meals, juices, fruit and vegetable preparations and dried foods. The most common active systems scavenge oxygen from the package or the product and may even be activated by an outside source such as UV light.

9.5.3 Intelligent packaging

Intelligent packaging can be defined as packaging that contains an external or internal indicator to provide information about aspects of the history of the package and/or the quality of the food. Intelligent packaging is an extension of the communication function of traditional packaging, and communicates information to the consumer based on its ability to sense, detect, or record external or internal changes in the product's environment. Intelligent packaging systems exist to monitor certain aspects of a food product and report information to the consumer. The purpose of the intelligent system could be to improve the quality or value of a product, to provide more convenience, or to provide tamper or theft resistance. Intelligent packaging can report the conditions on the outside of the package, or directly measure the quality of the food product inside the package. In order to measure product quality within the package there must be direct contact between the food product or headspace and the quality marker. In the end, an intelligent system should help the consumer in the decision making process to extend shelf life (thus minimizing food waste), enhance safety, improve quality, provide information, and warn of possible problems. Intelligent packaging is a great tool for monitoring possible abuse that has taken place during the food supply chain. Intelligent packaging may also be able to tell a consumer when a package has been tampered with.

Intelligent packaging systems monitor the condition of packaged foods to give information; packaging system that senses and informs. Intelligent packaging systems facilitate decision making, extend shelf life, enhance safety, improve quality, provide information, and warn about possible problems by their capability of carrying out intelligent functions (like detecting, sensing, recording, tracing, communicating, and applying scientific logic).

Examples are time-temperature indicators, ripeness indicators, biosensors (e.g. sensing volatile products of microbial growth), radio frequency identification (RFID), and gas sensors.

9.5.4 Biodegradable packaging film

These are plastics that will decompose in natural aerobic (composting) and anaerobic (landfill) environments. Biodegradation of plastics occurs when microorganisms metabolize the plastics to either assailable compounds or to humus-like materials that are less harmful to the environment. They may be composed of either bio-plastics, which are plastics whose components are derived from renewable raw materials, or petroleum-based plastics which contain additives. The main benefit is that these can completely metabolize to carbon dioxide (and water) and are more environment friendly, allowing complete recovery of large quantities of waste.

Biodegradable packaging films are one of the new environmental friendly trends for promoting sustainability in food industry. Apart from the general task of packaging material, protection, communication, convenience and containment,, biodegradable packaging films are capable of decaying through the action of living microorganisms. Thus introducing such materials in food processing fosters preserving the environment. The materials used for producing these films are limited to renewable sources.

9.5.5 Radio frequency identification

RFID is a technology that uses a wireless non-contact system that uses radio-frequency electromagnetic fields to transfer data from an electronic chip on its 'host' attached to an object, for the purposes of automatic identification and tracking. Data is stored on this chip and can then be read by wireless devices, called RFID readers. RFID tags are smarter than the traditional barcodes as the information on the micro-chip can be read automatically, at any distance, by another wireless machine. This makes RFID application easier and more efficient than barcodes.

RFID is applied to food packages for supply chain control and allowing food producers and retailers create full real time visibility of their supply chain. Earlier applications have been in livestock inspection and tracing. It makes goods tracking easier and cannot be easily replicated (like barcodes) and therefore increases the security of the product, capable of storing data up to 2 KB, whereas the bar code has the ability to read just 10-12 digits. It is more effective in harsh environments, where bar code labels have problems, allows automatic scanning and data logging, can be in read only or read-write. Shortcomings are the high cost and the difficulty for an RFID reader to read the information in case of liquid and metal products. The presence of mobile phone towers has been found to interfere with RFID radio waves too.

9.6 Conclusions

A variety of new technologies are in the focus of research and are of interest for food sector stakeholders. Many of these technologies are inherently promising. However, unless these meet the industry's safety, quality and sustainability benchmarks, they are 'doomed to failure' at lab or pilot scale. There are some do's and don'ts that are critical to bear in mind to ensure a technology's success in the food industry. For example, food safety is a given in the food industry while food quality serves as a competitive differentiator in the global market place. Therefore, any novel technology that enhances food safety should not compromise the inherent quality attributes that made the particular product successful in the market.

An ever-changing lifestyle and an increasing quality awareness of consumers living in prosperous societies, where they are more focused on what they eat and can choose from an enormous variety of foods, have created a very dynamic food sector requiring alternative concepts of food production. The efficiency and sustainability of food processing can directly be influenced by the choice of a suitable technology. Besides raw material quality, the selected technology and intensity of processing strongly influence important food attributes such as flavour, texture, appearance and nutritional composition. The potential of alternative food technologies was investigated already in the beginning of the 20th century. Since then significant scientific effort has been spent on evaluation of the impact of alternative technologies on quality attributes of different foods, as well as on microbial inactivation and enzymes. In parallel with studies on treatment effects, equipment design, process optimisation and machine development have developed tremendously. Following significant research and technological achievements on alternative food technologies with less harmful effects than conventional thermal treatments, some of these technologies were successfully introduced in the industry. These technologies are able to improve process efficiencies and food quality while at the same time delivering the same level of safety.

Many of the technologies mentioned in the chapter owe their emergence to consumer demands. Some of these technologies have enabled the food industry to develop new formulations and products that never existed before to meet consumer demands and requirements. The growing global trend towards 'fresh' and 'natural foods' has led to adoption of some mild and non-thermal processing technologies. The most advanced are HPP, PEF, OH and eBeam technologies. These technologies have contributed to the availability of nourishing, convenient, safe and high quality food products. eBeam technology in particular has enabled pushing the boundaries of what one can describe as high quality, nutritious, extended shelf-life, microbiologically safe foods with the ability to produce sterilized food items. However, although the scientists and industry mostly welcome progress in research and machine development, consumers are more cautious towards innovations in food they eat. One positive example is certainly HPP, where the majority of consumers willingly accepted the technology and the products produced by it. On the other hand, scientific advances in GMO and irradiation were rejected by a large majority of the consumers and some governments as well.

Recently, more and more food labels contain claims on what they are free from. A new trend 'from clean to clear label': clear and simple claims as well as maximum transparency, was identified as a number one trend in 2015. Technologies such as HDP or HPH have enabled food processors to adopt 'clean labelling'. In this way product information can support consumers to make more educated choices regarding a balanced diet and a healthy lifestyle. RFID technologies have made significant impacts on food traceability. However, many new technologies are still being pilot tested. Novel packaging approaches are still at their infancy and need further customization to cheapen their production and make them beneficially applicable commercially. However, no matter how elegant and justifiable a new technology appears, unless the technology makes a financial justification for adoption, the technology will languish at the drawing board or pilot scale. The cost of applying a new technology is one of the main drawbacks in adopting it unless it contributes to energy savings or emergence of more value added products. Similarly, strict local food regulations can also discourage the adoption of the new technology.

The food industry has enormous potential for innovation, but at the same time, it is rather slow and conservative. The food sector stakeholders have to be accurately informed about the pros and cons of each technology and have to be very careful when choosing the right technology for their product. Failure in doing so, coupled with hype can cause the rapid loss in trust. The applicability of a technology can be recognised based on different important parameters (while considering product-technology suitability), such as quality benefit, energy and costs requirements, potential environmental impact, waste handling, market and marketing opportunities. Therefore, attention paid to these parameters, naming of a novel technology, the terminologies used, the quality of information provided to the stakeholders can all play a major role in the successful transition from an experimental technology to a commercially viable technological solution.

References

Aganovic, K., Grauwet, T., Kebede, B.T., Toepfl, S., Heinz, V., Hendrickx, M. and Van Loey, A., 2014. Impact of different large scale pasteurisation technologies and refrigerated storage on the headspace fingerprint of tomato juice. Innovative Food Science and Emerging Technologies 26: 431-444.

Cannon, R.J.C., Hallman, G.J. and Blackburn, C., 2012. The pros and cons of using irradiation for phytosanitary treatments. Outlooks on Pest Management 23: 108-114.

Desrumaux, A. and Marcand, J., 2002. Formation of sunflower oil emulsions stabilized by whey proteins with high-pressure homogenization (up to 350 MPa): effect of pressure on emulsion characteristics. International Journal of Food Science and Technology 37: 263-269.

Donsi, F., Ferrari, G. and Maresca, P., 2009. High-pressure homogenization for food sanitization. In: Barbosa-Canovas, G.V., Mortimer, A., Lineback, D., Spiess, W., Buckle, K. and Colonna, P. (eds) Global issues in food science and technology. Academic Press, San Diego, CA, USA, pp. 309-352.

Floury, J., Bellettre, J., Legrand, J. and Desrumaux, A., 2004a. Analysis of a new type of high pressure homogeniser. A study of the flow pattern. Chemical Engineering Science 59: 843-853.

Floury, J., Legrand, J. and Desrumaux, A., 2004b. Analysis of a new type of high pressure homogeniser. Part B. Study of droplet break-up and recoalescence phenomena. Chemical Engineering Science 59: 1285-1294.

International Food Information Council (IFIC), 2014. The pulse of America's diet from beliefs to behaviors. International Food Information Council Foundation 2014. Food and Health Survey. Available at: https://tinyurl.com/ya9lq4yo.

Ivorra, A. and Rubinsky, B., 2010. Historical review of irreversible electroporation in medicine. In: Rubinsky, B. (ed.) Irreversible electroporation. Springer, Berlin Heidelberg, pp. 1-21.

Pillai, S.D. and Shayanfar, S., 2015. Electron bean pasteurization and complementary food processing technologies. Woodhead Publishing, Cambridge, UK.

Sebranek, J.G., Dikeman, M. and Devine, C.E., 2014. Irradiation. In: Dikeman, M. and Devine, C. (eds.) Encyclopedia of meat sciences, 2nd edition. Academic Press, Oxford, UK, pp. 140-144.

10. Forms of food distribution and trends in food retailing

B. Borusiak[*] and B. Pierański

Poznań University of Economics and Business, al. Niepodległości 10, 61-875 Poznań, Poland; barbara.borusiak@ue.poznan.pl

Abstract

The aim of this chapter is to present the various forms of food distribution, both direct and indirect. Since retailers now occupy a special position within indirect distribution channels, the discussion will focus on the formats of food retailing and the most important trends observed in this sector. Modern systems of food distribution are very diverse; however, polarization can be clearly seen. At one extreme, there are complex, often long, distribution channels for food produced on a mass scale, with many intermediaries, including very strong retailers, often conducting business on a global scale. In these channels one can observe a number of new phenomena associated with modern technologies. At the other extreme are simple, direct distribution channels for non-industrial food products, minimally-processed and produced by natural methods, which in the less prosperous regions of the world represent the most traditional and frequent type of channel. At the same time, the latter type also appears in economically developed countries as a manifestation of a return to nature and a search for food that is healthy, natural and unspoilt by chemical additives.

Keywords: distribution channels, store formats, trends in retailing, innovation

Klaus G. Grunert (ed.) **Consumer trends and new product opportunities in the food sector**
DOI 10.3920/978-90-8686-852-0_10, © Wageningen Academic Publishers 2017

10.1 Forms of food distribution

Distribution is the process of transfer of goods from the place of production to the place of consumption. It is one of the instruments of marketing, and its structure comprises two components: distribution channels and systems for the physical distribution of products. The former will be the focus in this chapter. A distribution channel is defined as a chain of separate economic entities which are involved in the movement of goods from the manufacturer to the consumer. Among them, the most important are trading companies, which are professional trade intermediaries, i.e. they buy goods for resale. The main categories of such companies are wholesalers and retailers. In the last 30-40 years, retailers have gained particular importance in food distribution channels, which will be discussed later in the chapter.

Food distribution channels can vary considerably in terms of the following features:
- Type of participants/involvement of intermediaries; on this basis direct (without intermediaries) and indirect channels can be distinguished.
- Channel length, defined as the number of intermediaries at the different levels of trade, i.e. buying, selling, wholesaling and retailing. Channels which have intermediaries at different levels are long distribution channels, while channels with intermediaries at only one level or no intermediaries at all are short channels.
- Channel width, defined as the number of intermediaries at each level: wide channels have many intermediaries at each level, as opposed to narrow channels.
- Degree of integration of channel participants: channels can be divided into integrated and non-integrated. The former include vertical contractual channels; among which a special type is administered channels, distinct because of the dominant position of one of the channel participants. Another type are channels integrated through capital (so-called own or corporate). In non-integrated channels (known as conventional), the participants are not tied by long-term contracts, and no other than market coordination mechanisms are used.

10.1.1 Forms of direct food distribution

Direct food distribution is organized by farmers. Its forms are presented in Table 10.1.

Table 10.1. Forms of direct food distribution (Collins, 2011).

On-farm	Off-farm
Gate/roadside stands	Vendor at farmers' markets
Farm markets/shops	Internet – online direct order
Pick-your-own operations	Vending machines
Community supported agriculture	

Gate/roadside stands

This is a direct form of food distribution involving little capital investment because gate/roadside stands can be as simple as a trailer or as elaborate as a small hut located somewhere close to the farm. In particular, this form can be used for selling seasonal products. Interestingly, in some countries, for example Denmark and the United Kingdom, this form of direct distribution can sometimes be entirely self-service (without, however, using vending machines). The buyer takes from the stand the products that they need and leaves the payment in the indicated place.

Farm markets/shops

These types of farm markets/shops range from seasonal to fully-functioning, year-round country stores offering consumers an alternative to the supermarket or regular grocery shops. Success often depends on good research into the local market, products and ways to develop customer loyalty. A drawback of this form is the need to devote time to it in addition to the farmer's core tasks.

Pick-your-own

These were very popular in the 1970s; those still remaining have become unique and different. Many have added the words 'edutainment' or 'agritourism', targeting school groups and families, or have expanded into special interest markets such as corporate picnics, film companies, etc. This form of distribution fits perfectly with the currently observed trend among consumers: collecting authentic experiences.

Community supported agriculture (on-farm pick-up)

In basic terms, community supported agriculture consists of a community of individuals who pledge support to a farm operation so that the farmland becomes, either legally or spiritually, a community farm, with the growers and consumers providing mutual support and sharing the risks and benefits of food production. The members or shareholders of the farm or garden pledge, in advance, to cover the anticipated costs of the farm's operation and the farmer's salary.

Farmers' markets

There are thousands of farmers' markets in many countries. They have regained popularity and are considered a community event. A full-season commitment to attend is required from the producer. Farmers' markets offer the customer variety, and quality customer service builds customer loyalty, bringing them back to the farmer's stall week after week.

Internet – online direct order

The use of the Internet and online direct ordering is increasing. This is an emerging form of food distribution which requires background research and development. It can build customer loyalty for products and is not limited in time, space or geography, and can take the form of a regular e-store (run by a farmer), offers on integrated Internet platforms or online direct order. In some cases, deals are based on long-term contracts between a farmer and a customer, which entail the customer being obliged to buy a given amount of goods. This form allows the farmer to plan production precisely and gives the customer access to fresh food.

Vending machines

These can be run by producers as well as intermediaries. Machines selling milk owned by farmers can be found for example in the Czech Republic, Slovakia, Austria, Italy and Poland.

10.1.2 Typology of food retail formats

The predominant types of food distribution channels in current business practice are indirect channels. Although they may vary considerably in length, in the last 30-40 years a significant growth has certainly been observed in the importance of retail enterprises. That is why retailers will be the focus of attention in this section.

A retail company concentrates its marketing activities on the end buyers, which means that goods are sold for the purpose of consumption, rather than, as in the case of wholesalers, for resale or for use in manufacturing. Retailers have at their disposal a large variety of distribution forms, which are known as retail formats.

As a starting point for the conceptualisation of a retail format, the essence of the product of a retail company must be explained. This product is a service, and the basis for the service is a set of goods, which is the range of products on offer. This means that a retailer's product is a service that is based on a set of products originating outside the retail sector. Essentially, it is the service element that provides the added value generated by a retail company (Dawson, 2000). In other words, the essence of a retailer's product is offering usability in the form of access to producers whose products are offered to the consumer in a convenient form in terms of configuration, location, time and volume, with favourable terms of service and payment.

A retailer's product, when considered structurally, has a very complex nature: it is defined by a number of components such as the form of the product assortment (in terms of dimensions such as depth, width and the criteria for its creation) and the conditions in which it is offered (location, time, presentation, range of services, price level, forms of payment, and how customers receive goods). This product meets two types of buyers' needs. The first type comprises needs that require a product from the product range of a shop to be satisfied. The other type of needs are to a large extent independent from the product assortment of a shop: social, aesthetic, educational, and needs connected with spending leisure time (Fiore and Kim, 2007).

The considerable complexity of a retailer's product means that, taking into account the actual transactions between a retailer and a customer, the product is highly individualised, which is typical of service products. On the other hand, it is essential to ensure mass (and also economical) access to goods, which prompts retailers to standardise products. An aggregated and standardised designation for the product/service of a commercial enterprise is the retail format, which is a specific manner for conducting sales, both store-based and non-store. Different retail formats are characterised by a specific configuration of such marketing strategy instruments as product and service assortment, pricing policy, the location and manner of offering products, as well as the type and configuration of the resources used (Reynolds *et al.*, 2007; Tiwari, 2009; Zentes *et al.*, 2007).

The classification of retail formats proposed here (Table 10.2) takes into account a multitude of criteria and is based on a set of marketing strategy instruments, the most important of which are product assortment for store-based formats (Guy, 1998), and the method of communicating with customers for non-store formats (Borusiak, 2014). The formats presented are of large or growing importance in economic practice, although they do not exhaust the set of existing formats.

Table 10.2. Typology of food retail formats.

Store-based food retail formats	Non-store food retail formats
Supermarkets	E-stores/e-commerce
Discount stores	M-stores/m-commerce
Hypermarkets/superstores	Online auctions
Convenience stores	Group buying
Small grocers	Cooperatives
Food/drink specialists	Markets/stalls/booths/pop-up stores
Concept stores	Itinerant trade
Forecourt stores	Vending machines sale

In food retailing, store-based formats have a dominant position. The most important of these are convenience stores, discount stores, hypermarkets and supermarkets. They generally trade in Fast Moving Consumer Goods (which, in addition to food products, also include a specific range of non-food products, depending on the format). Common features of these formats include the use of the self-service method of sales, as well as the fact that they most frequently function as chains, often on an international scale.

The supermarket format has a relatively long history. It first emerged in the United States, mainly as a result of introducing the self-service method of sales, which took place in 1916. This format is one of the most important innovations in the history of the retail trade, which, through a number of modifications, has given rise to many other formats. Definitions of a supermarket vary from country to country. Basically, it is a store with an area no larger than 2,500 m² which offers a relatively wide and deep range of food products as well as a range of selected, frequently purchased non-food products. The supermarket is a format that comprises stores that differ significantly in terms of price levels, the services offered and location.

The concept of a discount store emerged in the United States as an innovative variation of the supermarket, probably in the late 1930s. In Europe, such shops appeared in the 1960s, introduced by the German company Aldi. The concept of a discount store is based on the following assumptions (Barth, 2002):
- the product range is wide, but at the same time quite shallow; this means that the costs relating to stock rotation can be significantly reduced;
- goods are offered at lower prices than in other formats.

As a result of implementing the above principles, discount stores are characterised by the following features:
- ► the product assortment typically does not exceed 1000 items and consists largely of own brands;
- ► the sales area usually does not exceed 1000 m².

A trend that can currently be observed with regard to discount stores is an evolution of the format, which consists in moving away from an assortment largely made up of the most basic products offered at the lowest possible prices. Now, the product range often includes high-quality goods (e.g. wine, seafood), widely recognized global brands, as well as, at the other extreme, locally manufactured products from certified producers (e.g. in the Austrian Hofer network, owned by the German company Aldi Einkauf GmbH & Co.).

There is also a variation of the supermarket called a hypermarket, a retail establishment with a sales area exceeding 2,500 m² (in some countries the minimum area for a hypermarket is larger, for example in Germany it is 4,000 m² and in the UK 5,000 m² (McGoldrick, 1990)), and with a wide and deep product assortment (up to 100,000 items), in which food products may no longer be the dominant type. Hypermarkets are typically located in city suburbs and offer relatively low prices (Horská *et al.*, 2010). The concept of a hypermarket originated in France as a result of the activities of Marcel Fournier and the Defforey family. The new format was based on the principle: everything under one roof. The first Carrefour hypermarket was opened on 15th June 1963 in Sainte-Genevieve-des-Bois, a southern suburb of Paris (Cliquet, 1998). Hypermarkets operate almost exclusively as chain enterprises, so in many countries the scale of their operations is very large.

A convenience store is a format in which the most important feature is convenience for buyers; hence, such stores are often located in areas with a high customer flow volume, and are open 7 days a week, at least 16 hours a day. The level of prices in convenience stores is higher than in other food stores: customers pay higher margins for the convenience in terms of time and space, and the possibility of doing the shopping quickly. A precursor of the concept of the convenience store is believed to be an American company, The Southland Corporation, which started to conduct its business through the 7-Eleven and Quick Mart chains in 1927 (though initially the stores were not self-service). A significant development of this format, however, occurred in the early 1960s. In its operation the convenience store largely resembled a corner shop (a neighbourhood shop, traditional in format), which often functioned as an independent establishment, whose main characteristic was a good location, and whose potential competitive advantage was built on close ties with regular customers (Jones, 1986).

Non-store retail formats do not yet have a large share in the sales of food products. Many of them, however, are developing very dynamically. These certainly include e-commerce and its variant – m-commerce. E-commerce has great potential for growth because of the great ease of purchase that it offers: at any time, without leaving home, and with the possibility of comparing the prices of products. Barriers to the development of e-commerce include the scale of Internet access and the conditions of this access, as well as the terms of delivery, which involve the necessity to incur additional costs and a time lag between making a purchase and the delivery of the goods. In the case of food, an additional barrier may be a lack of confidence in the quality of the products that the customer was not able to inspect by themselves.

Another very interesting non-store food format seems to be cooperatives. They are defined as open, informal, ad hoc and voluntary non-profit associations of people which make collective purchases of food (Abramowski, 2012). It should be mentioned that the Internet is a significant factor enabling the formation, operation and development of cooperatives, being a medium via which the group members can consult with one another, inform one another about new sources of supplies, etc. *(Borusiak et al., 2015)*.

Cooperatives are a specific retail format that provides certain benefits to its members. One of them, undoubtedly, is related to the possibility of buying eco-friendly products at prices lower than market rates. The prices of products ordered through cooperatives amounts to 70-80% of the prices of similar products purchased in traditional stores (Oszczepalski, 2012). This is possible because of an innovative approach to business. Firstly, it manifests itself in a maximum shortening of the supply chain. Cooperatives purchase products directly from farms. On the one hand, this creates the possibility of controlling the quality of the products bought, and fosters establishing direct relationships with farmers. On the other hand, it helps to reduce supply costs because there are no intermediaries, and the goods are transported using resources belonging to members of the cooperative. An innovative approach to relationships with suppliers is also evident in the area of pricing. The aim of a cooperative, as mentioned earlier, is to sell products at relatively low prices, but without exerting pricing pressure on suppliers. On the contrary, by promoting the principle of sustainable development, cooperatives offer 'honest prices for honest work'. Reducing operating costs is evident not only in the area of supply. A cooperative, being an informal entity, does not incur costs connected with maintaining an executive board or management. In addition, the ad hoc nature of the operation helps to minimise the cost of maintaining the premises in which the cooperative meets. Finally, the exclusive reliance on the Internet (social media) for communication between members and ordering products reduces costs connected with the organization of the group.

In summary, the forms of food distribution are highly diversified. In particular, there is a large number of retail formats. These formats are subject to the life-cycle phenomenon: new ones emerge and older ones lose their significance. This phenomenon occurs to varying degrees in individual countries. Undoubtedly, however, one can observe a strong increase in the importance of retailers in food distribution. This chapter does not deal with issues related to wholesale trade. This is partly because the function of wholesalers is often taken over by large retailers. The trends currently observed in the retail sector are explored in the next section.

10.2 New trends in retailing

10.2.1 Growth of retailer power

The growing potential of retail companies and, consequently, their growing bargaining power over suppliers, which often makes them administrators of the entire supply chain, is a phenomenon that has been observed since the turn of the 1990s (Buzzel *et al.*, 1990; Messinger and Narasimhan, 1995).

The main cause for this phenomenon is the mechanism of economies of scale in the retail sector (Borusiak, 2004). This is based on the retail-specific structure of total costs, which consists of two main groups of costs: the purchase price value of goods sold (cost of purchase) and the cost of sales, including the value of fixed assets used, materials, salaries and other expenses associated with running a business. The proportion of purchase costs in the total costs is particularly significant, often exceeding 70% of total costs. Such a high share of purchase costs prompts retailers to seek savings in this area because such actions are more effective than reducing the cost of sales. Thus, it can be concluded that increasing the scale of retail operations is an action that directly affects the variable cost per unit, which is the unit price of purchased goods. This means that suppliers, in exchange for the opportunity to have a high turnover, are willing to sell products to large retail customers at lower prices. Increasing the scale of retail operations has become possible through employing the self-service method of selling, which, through the mechanism of impulse purchases, has significantly contributed to an increase in turnover; as did introducing large retail formats such as supermarkets, discount stores and hypermarkets (Appel, 1972; Gilchrist, 1953).

As a result, a number of powerful commercial enterprises operate in many developed markets, with turnovers that put them among the largest companies both on a global scale (for instance Walmart was in the first decade of the 21st century the largest company in the world in terms of turnover) and in individual markets (such as Jeronimo Martins in Poland that was one of the top five of all the companies in 2013-2014 operating in Poland). A very valuable asset that these companies have is direct relationships with end users, who seek to purchase products locally, near their homes. Hence, retail enterprises form chains that operate in many local markets. Producers, on the other hand, do not generally have a direct relationship with consumers, or the scale of these relationships is much smaller. In view of the fact that as many goods do not need to be manufactured in the place where there is a demand for them, whereas the services provided by the retail sector are more strongly connected to a given location, retailers have gained an advantage over many suppliers in terms of market access. As a result, the pressure exerted by retailers on manufacturers has proved effective: manufacturers, fearing a loss of sales opportunities, accept terms of cooperation which are not always beneficial to them, especially with regard to types of products, prices, delivery conditions, etc.

There is a view, perfectly illustrated by the phrase 'retailer as gatekeeper', according to which in many cases retailers play a decisive role in the distribution channels of suppliers, because they make a selection of the goods offered by suppliers and in this way determine the range of products on offer, which can influence demand and the buying behaviour of consumers (Gilbert, 1999). This pattern is additionally stimulated by the following phenomena (Stern *et al.,* 2002):

► the amount of shelf space does not increase as fast as the number of new products;
► the retail market is becoming consolidated;
► the number of brick-and-mortar outlets is declining;
► existing technologies make it possible to very precisely determine the sales of specific products (information asymmetry: the retailer has more information about sales than the manufacturer);
► there is mounting pressure to increase resource productivity in the retail trade.

The phenomenon of the increasing power of retailers and the pressure they exert on suppliers entail a number of consequences. The most important of them seems to be maintaining a relatively low level of prices for so-called convenience products (Fast Moving Consumer Goods), including food. This, however, forces manufacturers to constantly look for methods to reduce costs (Watts and Goodman, 1997). In the case of food this happens through striving to increase economies of scale: operating on a large scale at the stage of sourcing raw materials, of both plant and animal origin, and at the processing stage (Opara, 2003). Another method for reducing costs is a disjunction of the value chain, which consists in considerable geographical fragmentation of some processes, for example food processing (Flejterski, 2011). This means that the labour-intensive stages of a process, e.g. fish processing, are performed in countries where labour costs are low. Also, new food production technologies are introduced, contributing to lowering costs. Such actions, however, cannot but have an effect on the quality of the food and the environment in which it is produced. Industrial food, which is highly processed and has a long shelf-life, is perceived in many affluent countries as food of inferior quality (Goodman, 1999). As a result, there is growing interest among consumers in organic, minimally processed food obtained from local suppliers (Murdoch and Miele, 1999).

In addition to the above-mentioned effects of the rising power of retailers, one can also mention others, such as introducing own-label products into stores. This practice makes it possible for retailers to have higher margins as well as building customer loyalty (by offering unique goods). Food is a category of products in which private labels began to appear in the late 19th century. A forerunner in this field was the British company J. Sainsbury, which in 1882 introduced smoked bacon under its own brand (Spyra, 2007). At the end of the 20th century, own brands already had a significant market share in many countries around the world (Lubańska, 2011) and this represented a serious threat to manufacturers, who were forced to compete with retail chains, and in the case of the manufacturers of own-label brands – with one another. In the past, private labels were used by retailers for the cheapest products in a given category. Nowadays, retailers offer portfolios of brands, which contain economy brands, 'value for money' brands, as well as premium and even luxury brands. Under their own brands, retailers offer organic and Fairtrade certified products (Lubańska, 2011). In a sense, this is proof of the possibility of separating the identity of the goods produced from the identity of the manufacturer. The identity of a product can be created by the seller, who in this case acts as the guarantor of the quality of the goods. But in order to do so, the retailer must build their own credibility, preferably resulting from a number of positive experiences by the customer, or from the positive experiences of a number of customers.

An important consequence of large scale operations by retailers is the opportunity to introduce modern technological solutions which, although characterized by relatively high initial cost, enable retailers to improve the quality of their services and reduce the cost of sales.

10.2.2 Modern technologies in the retail sector

The main purpose of introducing modern technological solutions in the retail sector is to enhance the shopping experience of customers. Such a conclusion can be drawn based on an analysis of survey results, according to which this aim is the most important for 70% of retailers. Next in the order of importance are such aims as lowering operating costs and gaining a competitive advantage (The 21st Century Store: the Search for Relevance, 2011). Ensuring the desired shopping experience requires integrating modern technologies with traditional merchandising techniques so that the impact on the customer will be multilateral, multifaceted, or, as it sometimes referred to, multi-channel (Dubbs, 2015). On the other hand, enhancing the experience of customers becomes inextricably linked with customizing their contact with retail establishments. In this area, technology is expected to make customers perceive the process of shopping as exceptional, unique, and tailored to the specific needs of individual clients. And although it may seem that the possibility of using new technologies by grocery stores, due to the nature of the products offered, is limited and less effective (for example in comparison to shops selling clothing), they do in fact introduce aspects of personalization to the process of purchasing food products.

It can be assumed that a fundamental role in terms of adopting the latest technological solutions by the retail sector was played by mobile technologies, specifically the spread of mobile phones that can connect to the Internet or communicate with other devices using Bluetooth technology. A growing number of buyers who have such phones have made it possible for retailers to employ technologies that are likely to play a significant role in shaping the future of this sector. These technologies include communication technologies using broadcasting devices (Beacon technology) and software for creating augmented reality (AR).

In order to perform its task, Beacon[3] technology requires the installation of a special application for smartphones (Lalik, 2014). Such applications scan the surroundings, searching for signals from specific beacons. When a beacon is located, a signal sent via Bluetooth activates a specific action in the application. Usually it is a personalized message encouraging the customer to visit a nearby store and purchase specific products. In addition to communicating personalized offers, beacons can be used in a number of other ways: from sending additional product information (e.g. potential allergens in a product) that is displayed on the screen, to navigating the customer around the shop. Through activating a specific action in the application (e.g. issuing voice commands about the direction of movement), the beacons installed inside a store can direct the customer to the place where a certain product can be found. Beacons also make it possible to display personalized messages on screens located in the shop window and inside the store (Satilli, 2015). This concept of using screens that display, for example, information, videos or advertising, the content and form of which can be changed in real time depending on the context, is called digital signage. A beacon installed, for example, at the entrance to a store, having connected with the mobile phone (with the appropriate application) of a passing customer, can display a personalized advertisement of a product on a screen installed in the shop window. In a less personalized version, although equally or even more technologically advanced, a camera installed in the screen, using the appropriate software, is able to recognize, for example, the gender and age of a customer. Based on this information, a specific advertising message is broadcast.

---

3   A beacon is a very small, lightweight device which can be installed on the wall or ceiling of a shop. Its task is to transmit signals. The transmission range is determined according to need: beacons can transmit signals over long distances (hundreds of metres), encouraging customers to visit the shop; or over short distances (up to one metre), informing customers about a specific product displayed nearby.

However, there is a risk that the advertised products may be stereotypically associated with a particular gender and/or age, for example beer for men.

The other technology mentioned above is AR. This is defined as a technology which 'imposes' a computer-generated image onto the real world, thus providing the user with an enhanced perception of reality. In other words, AR is a complementation of real objects with information from digital sources (Daszkiewicz, 2013). Reality is enriched by so-called markers put on various items, for example shelves or specific products. These markers, recorded by a camera with the right software, generate in real time the content assigned to them, which will be visible on the screen of, for example, a computer (http:// www.alphavision.pl/rzeczywistosc-rozszerzona-augmented-reality-.html). This content can be a picture, text, video or music. The devices which enable the use of AR in retail are smartphones. They have a built-in camera, a screen, and enough memory to store the software (applications). These applications are commissioned by retailers who want to enhance the buying experience of consumers. In food retailing this technology can be used in two areas: helping customers to locate specific items and providing additional information about products. In both cases, AR will reveal itself when looking at the environment (e.g. the coffee aisle in a grocery store) through a smartphone. Appropriately positioned markers can then flag the products of a specific manufacturer, free of certain substances (e.g. gluten free), coming from local suppliers, etc. In turn, the markers placed on the packaging of specific foods in AR can take the form of a detailed description of the product's composition, a list of recommended complementary products, or recipes in which the product is a key ingredient. In more advanced (personalized) versions of AR, one can imagine that the kind of dishes that the recipes will suggest or the amounts of the necessary ingredients will be adjusted according to the size of the customer's family or their lifestyle. Another advantage of AR is that because of the markers placed on products (products themselves can also be markers), information about them will also be available to the customer after doing the shopping; for example when they want to use the products at home.

AR, being a new technology, is still in the initial stage of use. Therefore, at present it is difficult to judge how widely it will be accepted and used by customers, and how profitable it will be for retailers. To date, there are no comprehensive and reliable studies diagnosing issues related to the impact of AR on customer satisfaction, volume of purchases, or retailers' revenues.

The prevalence of mobile phones also allows retailers to contact customers using technology that is less sophisticated than Beacon or AR. Retailers quite commonly use 'ordinary' applications which, when installed on a smartphone, show a map of the store with the precise location of individual product categories and make it possible to send personalized promotional coupons to individual customers. An interesting application was created by Tesco. The residents of Seoul in South Korea were able to scan the QR codes of products that were displayed on the wall of a subway station. After selecting (by scanning the codes) the products, the customer was able to pay for them using the application. Then the 'actual' products were delivered from the Tesco warehouse to the customer's home.

The technologies discussed above can substitute each other (e.g. for indicating the location of specific items inside the store), or complement one another (AR offers the possibility of conveying to the customer much more extensive product information in comparison to the information transmitted to mobile phones by beacons). Therefore, regardless of the type and number of technologies employed, it is crucial to integrate them. Only appropriate synchronization will ensure that they will meet the expectations of both customers and retailers.

10.2.3 Personalisation of retail offers

The personalization of retail offers is currently regarded by many experts in the field of retail trade as the highest priority and at the same time as the biggest challenge for management (Grant, 2015). Despite the attention given to the phenomenon of personalization, it has not been clearly defined. This is due to the fact that the term covers a very wide spectrum of concepts related to, among other things, building relationships with buyers, collecting and analysing market information, as well as designing production processes (Kemp, 2001). Nevertheless, a review of the different definitions makes it possible to identify the undisputed constituent feature of personalization, which is adjusting the company's offering to the individual needs of each client. This is mentioned by Hanson (2000), who states that personalization is a form of product differentiation that enables meeting the individual needs of customers. A similar view is expressed by Peppers (1999), according to whom the essence of personalization is matching the characteristics of a product to the individual needs of customers so that they can enjoy greater convenience (of using the product), lower costs, and other benefits. Finally, personalization is seen as a company's ability to recognize the needs of its customers and treating them on an individual basis (Imhoff *et al.*, 2001). At this point it should be noted, however, that in the literature there is no agreement as regards defining the concept of personalization and related concepts such as customization, prosumption, co-creation, etc. Opinions in this respect are divergent. According to some researchers, customization is a concept distinct from personalization, the distinction being based on the type of information used to create a specific offer (Szymkowiak, 2014). According to others, customization is included in personalization, being a separate component of the process (Hanson, 2000; Imhoff *et al.,* 2001). According to still other researchers, these two concepts are synonymous and can be used interchangeably (Peppers *et al.,* 1999).

Undoubtedly, the cause of the increased interest in and use of personalization is the rapid development of information technology, which makes it possible to collect, process and analyse data to identify and study customer needs (McCarthy, 2004). This is because personalization is a process whose inherent element is continuously obtaining information about buyers. This information can be obtained either through analysing the market behaviour of the company's customers, or acquiring data from external sources such as reports. Modern information technologies make it possible to collect enormous amounts of diverse data, known as Big Data, and appropriate software helps to adequately analyse these data (Borusiak *et al.*, 2015).

The personalization of retail offers has a very long history. It is connected to the earliest stages of the sector's operation when merchants served customers who they knew very well; thus, they had sufficient information to prepare dedicated offers. It was only in the 20th century, with the introduction of self-service stores, that retailing underwent a process of massification. Now, however, there is a strong need for, and at the same time the possibility of, personalization, which does not exclude catering for the mass customer. Thus, the main prerequisites of personalization are the following:
- customer willingness to share data about their shopping habits;
- direct access to data relating to the buying behaviour of customers;
- advanced technological possibilities for collecting and processing information, leading to creating customer profiles.

When comparing online and traditional stores, it seems that the former have a much greater range of personalization options. For one, this is because such stores obtain a great deal of information about their customers every time they make an online purchase. Knowing a customer's personal data, address and shopping history, they can create individual profiles of each customer. In traditional shops, on the other hand, especially self-service ones oriented to mass sales, the customer can remain completely anonymous. Thus, in this case the possibilities for personalizing the offer seem to be limited. That is why stores selling food products face problems related to the personalization of their offers. As research shows, the vast majority of European consumers buy their food products in traditional stores (that do not operate online); therefore, in 2015, for example, only between 4 and 14% of the population of European countries bought food products on the Internet. A marked exception is the UK, where the share was 29% (http://tinyurl.com/yc2s7wsz). This state of affairs may well have been the reason for the increased emergence of retailers with both online and bricks-and-mortar stores. Possessing an extensive range of formats, brick&click establishments have on the one hand, direct access to customers, representing the traditional understanding of personalization; and on the other hand, they have access to information about different types of customer behaviour relating to both shopping and other online activities. Exploiting these benefits requires an integration of the different formats in a less (multi-channelling) or more (omni-channelling) sophisticated form.

The key element of a personalized offer, in accordance with the definitions quoted above, is matching the characteristics of a product to the individual needs of buyers, which in the retail sector is implemented in a different way than in the manufacturing sector. Retail companies individualize their offer not by modifying the physical form of individual goods, but by providing the consumer with ready-made purchasing combinations. This means that the customer does not need to view (browse) all the products from the offered range, but is presented only with a section that corresponds to their current and/or potential needs. In view of the currently frequently observed behaviour of analysis paralysis, being a consequence of the multitude of options to choose from (Schwartz, 2013), such a practice leads to shortening the transaction time.

10.2.4 Multichannel retailing

Creating portfolios of formats by retailers is not a new phenomenon. For many years, retail companies, aiming to increase their sales revenues, have created new chains of stores in new formats. Commonly, these formats are similar to existing ones in terms of product assortment, the retail technologies used, or the characteristics of the customer groups (Borusiak, 2008). However, from the customers' point of view, such chains often function as autonomous entities (Herhausen *et al.,* 2015); moreover, the different names of the chains suggest that they compete with each other.

Multi-channelling in the retail trade is a new quality because it is a model of doing business in which the retailer uses two or three integrated retail formats, including at least one store-based and one non-store format (Kucharska, 2015). In particular, those retailers who rely on store-based formats recognize the need for using online formats in order to reach customers with their offers (Neslin and Shankar, 2009). This is connected with the increasing virtualization of consumer behaviour, particularly in the case of young consumers, manifested in the following trends (Kucharska, 2014):

- an overlap between the real and the virtual world; an increase in the number of digital products and digital distribution channels; an increase in digital communication and social networking; solutions such as the Internet of Things and AR;
- 'immersion' in the world of modern communication technologies; a constant presence on social networks, forums, etc.;
- perception of the world as a game as a result of intense involvement in online gaming; young consumers seek to meet their needs through participation in the design of new products and activities on Internet forums;

- expectation of instant gratification, constant access to shops and shopping opportunities;
- cocooning – people retreat into their homes; with work, shopping and social contacts being done online;
- the ROPO effect (research online, purchase offline), also referred to as webrooming.

The integration of formats (sales channels), which is the basis of multichannel retailing, primarily consists in openly communicating the fact of being part of a certain business entity or group of companies, which is a direct response to the last of the above-mentioned trends. A multichannel retailer does not see the ROPO effect as a threat, but as a natural phenomenon that must be faced. Integration of sales channels is defined as the degree to which different formats interact (Bendoly *et al.*, 2005). This means that one format contains functionalities serving the other format, such as a store locator on the online store's website, the possibility of obtaining information about the availability of goods in the offline shops from the Internet shop's website, or the possibility of browsing the online product catalogue in the traditional shop. The benefits of multichannel retailing include raising the value of the offering (Gallino and Moreno, 2014) and preventing the frustration of customers (Gulati and Garino, 2000). However, multichannel retailing is also connected with a number of dangers, such as the consequences of insufficient integration and the cannibalisation of formats/channels (Herhausen *et al.*, 2015).

In the case of food retailing, an important consequence of a multichannel strategy is the impact on the level of risk that customers associate with purchasing food on the Internet. Experience (also from the Polish market) shows that a retailer who has a chain of bricks-and-mortar stores with high quality products is more effective in developing an online channel for the sales of food, particularly fresh food, compared to a retailer who uses exclusively an e-commerce format.

A more advanced form of sales channel integration is omni-channel retailing, which involves exploiting additional sales channels, in particular mobile commerce, based on using smartphones with appropriate applications. Omni-channel retailing leads to the disappearance of boundaries between retail formats that are used interchangeably at the stage of searching for information, completing the transaction, as well as post-transaction services. For example, customers may simultaneously look for information about a product in the store and use applications on smartphones or tablets (which may be supplied by the shop) that help them find a better deal elsewhere. Or a customer can place an order via an application on their smartphone, but collect the goods in person from the store: there could be numerous combinations in the ways to complete transactions (Verhoef *et al.*, 2015). Although omni-channel retailing may be a considerable organizational challenge, it offers substantial opportunities for gaining a competitive advantage, which is connected, among other things, with the previously discussed possibility of personalizing the offers while taking into account the context of time and space.

10.2.5 Corporate social responsibility in the retail sector

Corporate social responsibility (CSR) is a concept according to which companies voluntarily assume responsibility for the economic, legal and ethical consequences of their activities; engage in charitable activities directed at customers, suppliers, employees and local communities; as well as protecting the environment (Carroll, 1999). It seems that in the case of food retailers, a measure that to the greatest extent meets the objectives of CSR is striving to shorten supply chains. This shortening is based on two interrelated activities. The first of these is eliminating intermediaries; the second is reducing the distance over which food is transported from the place of production to the shops. The second aspect is related to a reduction in the so-called food-miles indicator. It represents the energy consumption involved in the transport of food from the place of production to the place of consumption. Most commonly, this energy consumption refers to the amount of fuel consumed by the means of transport used (trucks, trains, ships, airplanes) (Hill, 2008). Thus, the food-miles indicator is closely related to the footprint indicator, which measures the negative impact of transport on the environment.

Obviously, these actions can only apply to those products that come from local livestock or crop farming establishments (located at relatively small distances from the points of sale). Purchasing products locally or regionally by retailers has a positive impact on a number of stakeholders listed in the definitions of CSR. These include the suppliers, the local community, and consumers. Finally, a reduction in food miles undeniably has a positive impact on the environment.

According to the original intention, the CSR concept was connected with the promotion of suppliers from distant countries; ensuring appropriate manufacturing conditions and fair wages, employing only adults in production, as well as preventing the devastation of the environment (Nestorowicz and Stefańska, 2012). Today, however, a special position in CSR activities is assigned to local suppliers. Purchasing food from local farms and processing and manufacturing plants translates into jobs for members of the local community and supports its development. Supporting local suppliers seems to be particularly important because for many years they were systematically eliminated from the market by global corporations purchasing products from distant countries. In 1870, in one of the states in the USA (Iowa), 100% of consumed apples came from local farmers, whereas in 1999 only 15% did (Pirog *et al.*, 2001).

The concept of CSR in the strategies of retailers towards suppliers is thus manifested in supporting local producers, developing local job markets to enhance their potential and prosperity, as well as ensuring for themselves a privileged position as a preferred shopping venue among consumers (Stefańska, 2014).

Undoubtedly, offering local food produce also has a positive influence on consumers. First of all, such food is fresh and rich in nutrients. What is more, it is not treated with those chemicals necessary for the food to withstand a few days' (or even a few weeks') transport and storage (Hill, 2008). In addition, considering that the tasks of CSR include striving to strengthen the local community, offering local products helps to create some very positive attitudes, such as the sensitivity of consumers to the origin of the food that they purchase. In addition, consumers have the opportunity to support local producers.

As mentioned above, offering local products also has a positive impact on the environment. According to estimates, the purchase of food products by retailers through traditional supply channels (from distant countries; thus requiring long transport lines that generate high levels in the food-miles indicator) leads to as much as a 17-fold increase in demand for fuel, which means a 17-fold increase in CO_2 emissions to the atmosphere in comparison to products purchased from local suppliers (Pirog *et al.,* 2001).

References

Abramowski, E., 2012. Zasada ekonomiczna kooperatywy spożywców. Available at: https://tinyurl.com/y7rld8ns.

Appel, D., 1972. The supermarket: early development of an institutional innovation. Journal of Retailing 48: 39-53.

Barth, K., 2002. The changing role of German hard discount store formats. European Retail Digest 36: 41-50.

Bendoly, E., Blocher, J.D., Bretthauer, K.M., Krishnan, S. and Venkataramanan, M.A., 2005. Online/in-store integration and customer retention. Journal of Service Research 7: 313-327.

Borusiak, B., 2004. Skala działania jako czynnik konkurencyjności przedsiębiorstw handlu detalicznego. Handel Wewnętrzny 4-5: 27-32.

Borusiak, B., 2008. Modele wzrostu przedsiębiorstw handlu detalicznego. Wydawnictwo Akademii Ekonomicznej w Poznaniu, Poznań, Poland.

Borusiak, B., 2014. The mechanisms for the emergence and evolution of retail formats. In: Musso, F. and Druica, E. (eds.) Handbook of research on retailer-consumer relationship development. IGI Global, Hershey, Derry Township, PA, pp. 108-126.

Borusiak, B., Pierański, B., Romanowski, R. and Strykowski, S., 2015. Automatyzacja personalizacji reklamy internetowej. Marketing i Rynek 3: 36-43.

Buzzell, R.D., Quelch J.A. and Salmon, W.J., 1990. The costly bargain of trade promotion. Harvard Business Review March-April: 141-149.

Carroll, A., 1999. Corporate social responsibility. Evolution of a definitional construct. Business and Society 38: 268-295.

Cliquet, G., 1998. Integration and territory coverage of the hypermarket industry in France: a relative entropy measure. International Review of Retail, Distribution and Consumer Research 8: 205-224.

Collins, D., 2011. Direct farm marketing in Ontario – A primer. Ministry of Agriculture, Food and Rural Affairs Fact Sheet. Available at: https://tinyurl.com/ycafmhpq.

Daszkiewicz, K., 2013. Augmented reality – Nowy wymiar rzeczywistości. Available at: https://tinyurl.com/y7jd9mg9.

Dawson, J., 2000. Viewpoint: retailer power, manufacturer power, competition and some questions of economic analysis. International Journal of Retail and Distribution Management 28: 5-8.

Dubbs, C., 2015. Top 2016 marketing trends and why you need interactivity. Available at: https://tinyurl.com/yaz3c7wd.

Fiore, A.M. and Kim, J., 2007. An integrative framework capturing experiential and utilitarian shopping experience. International Journal of Retail and Distribution Management 35: 421-442.

Flejterski, S., 2011. Proglobaliści, antyglobaliści i alterglobaliści. Spojrzenie z punktu widzenia zarządzania finansami. Zeszyty Naukowe Uniwersytetu Szczecińskiego 640: 243-254.

Gallino, S. and Moreno, A., 2014. Integration of online and offline channels in retail: the impact of sharing reliable inventory availability information. Management Science 60: 1434-1451.

Gilbert, D., 1999. Retail marketing management. Prentice Hall, Upper Saddle River, NJ, USA.

Gilchrist, F.W., 1953. The discount house. Journal of Marketing 17(3): 267-272.

Goodman, D., 1999. Agro-food studies in the 'age of ecology': nature, corporeality, bio-politics. Sociologia Ruralis 39: 17-38.

Grant, G., 2015. Personalization: retail marketing's priority for 2015. Available at: https://tinyurl.com/y7c6b3g8.

Gulati, R. and Garino, J., 2000. Get the right mix of bricks and clicks. Harvard Business Review 78(3): 107-114.

Guy, C.M., 1998. Classifications of retail stores and shopping centers: some methodological issues. GeoJournal 45: 255-264.

Hanson, W., 2000. Principals of internet marketing. South West, Cincinnati, OH, USA.

Herhausen, D., Binder, J., Schoegel, M. and Herrmann, A., 2015. Integrating bricks with clicks: retailer-level and channel-level outcomes of online-offline channel integration. Journal of Retailing 91: 309-325.

Hill, H., 2008. Food miles: background and marketing. Available at: https://tinyurl.com/ycq3yp9j.

Horská, E., Nagyova, L. and Rovny, P., 2010. Merchandising a event marketing. Slovak University of Agriculture, Nitra, Slovakia. Available at:

https://tinyurl.com/yc2s7wsz.

Imhoff, C., Loftis, L. and Geiger, J., 2001. Building the customer-centric enterprise, data warehousing techniques for supporting customer relationship management. John Wiley and Sons, New York, NY, USA.

Jones, P., 1986. The development of convenience stores in Britain. Service Industries Journal 6: 390-399.

Kemp, T., 2001. Personalization isn't a product. Internet Week 864: 1.

Kucharska, B., 2014. Innowacje w handlu detalicznym w kreowaniu wartości dla klienta. Uniwersytet Ekonomiczny w Katowicach, Katowice, Poland.

Kucharska, B., 2015. Multichannel retailing – Consumer's and retailer's perspective. Handel wewnętrzny w Polsce 2010-2015. IBRKiK, Warszawa, Poland.

Lalik, E., 2014. O co chodzi z tymi beaconami? Available at: https://tinyurl.com/yc8sogbs.

Lubańska, A., 2011. Znaczenie marek własnych sieci handlowych w Polsce. Ekonomika i Organizacja Gospodarki Żywnościowej (87). Available at: https://tinyurl.com/ychrv62x.

McCarthy, I.P., 2004. Special issue editorial: the what, why and how of mass customization. Production Planning and Control 15: 347-351.

McGoldrick, P.J., 1990. Retail marketing. McGraw-Hill Book Company, London, UK.

Messinger, P.R. and Narasimhan, C., 1995. Has power shifted in the grocery channel? Marketing Science 14: 189-223.

Murdoch, J. and Miele, M., 1999. Back to nature: changing worlds of production in the food sector. Sociologia Ruralis 39: 465-483.

Neslin, S.A. and Shankar, V., 2009. Key issues in multichannel customer management: current knowledge and future directions. Journal of Interactive Marketing 23: 70-81.

Nestorowicz, R. and Stefańska, M., 2013. Tendencje w komunikacji przedsiębiorstw handlu żywnością z konsumentami. In: Handel Wewnętrzny w Polsce. Rynek artykułów żywnościowych, IBRKK, Warszawa, Poland, pp. 192-205.

Opara, L.U., 2003. Traceability in agriculture and food supply chain: a review of basic concepts, technological implications and future prospects. Food, Agriculture and Environment 1: 101-106.

Oszczepalski, B., 2012. Kooperatywy przeciwko systemowi. Available at: https://tinyurl.com/yb6t59x3.

Peppers, D., Rogers, M. and Dorf, B., 1999. The one to one fieldbook: the complete toolkit for implementing a 1 to 1 marketing program. Double Day, New York, NY, USA.

Pirog, R., Van Pelt, T., Enshayan, K. and Cook, E., 2001. Food fuel, and freeways: an Iowa perspective on how far food travels, fuel usage, and greenhouse gas emissions. Leopold Center for Sustainable Agriculture. Available at: https://tinyurl.com/yb8c7lr9.

Reynolds, J., Howard, E., Cuthbertson, C. and Hristov, L., 2007. Perspectives on retail format innovation: relating theory and practice. International Journal of Retail and Distribution Management 35: 647-660.

Satilli, J., 2015. How digital signage is leading retail's 'Internet of Things'. Available at: https://tinyurl.com/y8mrkhbm.

Schwartz, B., 2013. Paradoks wyboru. PWN, Warszawa, Poland.

Spyra, Z., 2007. Kanały dystrybucji: kształtowanie relacji. Wydawnictwo CeDeWu, Warszawa, Poland.

Stefańska, M., 2014. Postrzeganie produktów pochodzących od lokalnych dostawców w kontekście etnocentryzmu i koncepcji CSR – Perspektywa nabywców. Marketing i Rynek 6: 713-727.

Stern, L.W., El-Ansary, A.I. and Coughlan, A.T., 2002. Kanały marketingowe. Wydawnictwo Naukowe PWN, Warszawa, Poland.

Szymkowiak, A., 2014. E-kastomizacja produktów – Wykorzystanie narzędzi online w e-commerce, Marketing i Rynek 8: 933-938.

The 21st Century Store: the Search for Relevance, 2011. Available at: http://tinyurl.com/yc524c8k.

Tiwari, R.S., 2009. Retail management. Retail concepts and practices, Global Media Mumbai, Mumbai, India.

Verhoef, P.C., Kannan, P.K. and Inman, J.J., 2015. From multi-channel retailing to omni-channel retailing. Journal of Retailing 91: 174-181.

Watts, M. and Goodman, D., 1997. Agrarian questions. In: Goodman, D. and Watts, M. (eds.) Globalizing food: agrarian and global restructuring. Routledge, London, UK.

Zentes, J., Morschett, D. and Schramm-Klein, H., 2007. Strategic retail management. Text and International Cases, Gabler, Wiesbaden, Germany.

11. Consumer segments in an international context

J. Paluchová

Department of Marketing and Trade, Faculty of Economics and Management, Slovak University of Agriculture, Tr. A. Hlinku 2, 949 76, Nitra, Slovakia; johana.paluchova@gmail.com

Abstract

In order to understand and interpret consumer liking, it is useful to identify commonalities in liking patterns, so that consumer responses can be clustered into groups. Segmentation, targeting, and positioning enable producers to avoid head-on competition in the marketplace by differentiating their offerings on the basis of such features as price, styling, packaging, promotional appeal, method of distribution, and level of service. The focus of the marketing concept is for marketers to know consumers' needs, and to secure, as accurately as possible, a picture of their likely future needs. In international market segmentation, consumers are segmented across countries and identifying groups of consumer in different countries that prefer similar attributes and represent similar buying behaviour is the essence of international segmentation. In this chapter, the strategic framework of segmentation is described in connection to the differences in consumer behaviour, in a European and world perspective, from different views such as cultural, geographical, demographical, and psychological differences. A segmentation strategy begins by selecting the base representing the core attributes of a group of existing or potential customers. This forms the basis for consumer segmentation.

Keywords: consumer segments, food marketing, market, segmentation criteria, target marketing

11.1 Introduction

Segmentation, targeting and positioning are the processes of dividing the total heterogeneous market into relatively distinct homogeneous sub-groups of consumers with common needs or characteristics and selecting one or more segments to target, because they have similar product/service needs (Batra and Kazmi, 2009). Market segmentation, strategic targeting, and product/service positioning are the key elements of marketing of consumer goods (Schiffman and Kanuk, 2010). Segmenting markets is defined as the identification of individuals or organizations with similar characteristics that have significant implications for the development of marketing strategy (Ürgeová and Horská, 2009). The purpose of segmentation is to identify differences in behaviour that have implications for marketing decisions, behavioural variables, such as benefits sought from the product and buying patterns (Jobber and Fahy, 2006). The strategy of segmentation presupposes that among the universe of consumers there are identifiable groups with characteristic special needs (Raju and Xardel, 2009). Segmentation divides a heterogeneous marketplace into smaller and more manageable homogenous components (Horská *et al.*, 2011). These smaller market segments can be targeted with more personally relevant positioning strategies that have greater appeal to individuals within the group. Segmentation variables that have been developed include demographic characteristics of the consumer, benefits sought by the consumer, and behavioural measures of the consumer (Reynolds and Olson, 2001).

11.2 Market segmentation strategies

A marketer will need to decide which strategy is best for a given product or service. Sometimes the best option arises from using different strategies in conjunction. The approaches to segmentation result from answers to the following questions where, who, why and how? (Kubicová, 2013). Figure 11.1 shows a diagram of four basic segmentation strategies distinguished in international or global marketing.

Another term for undifferentiated marketing is mass marketing. Rather than developing different marketing strategies for various segments, or even different products for different groups, undifferentiated marketing attempts to reach all potential buyers using one marketing strategy. In this way, undifferentiated marketing treats all segments of the population the same, and the strategy is to use one approach that aims to appeal to as many people as possible (Kotler and Armstrong, 2004; Schiffman and Kanuk, 2010). Coca Cola for most of its history produced just one soft-drink; now, Coca Cola regularly launches new products every year in many markets.

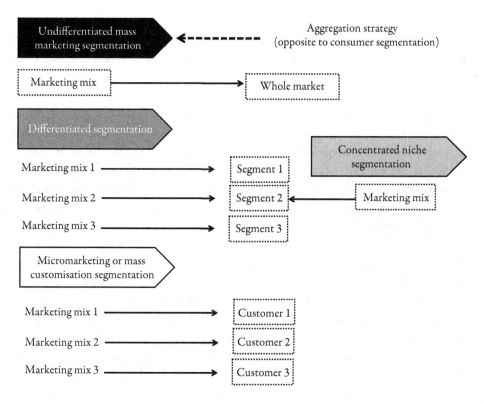

Figure 11.1. Four market segmentation strategies (adapted from Jobber and Fahy, 2006 and Kardes *et al.*, 2014).

Traditional segmentation strategies are called differentiated segmentation. A firm targets several market segments with the goal of having a strong position and share of the market in each segment. Trying to sell to two or more specific market segments creates a differentiated marketing strategy. This allows sellers to increase potential shopper base, sales, revenues and profits (Solomon, 2011). A company that runs three family-style Italian restaurants can create a differentiated segmentation strategy by opening a Mexican restaurant and/or a sports bar to pursue those market segments.

A special form of differentiated segmentation, called concentrated niche segmentation, is where a brand attempts to capture a large share of a smaller niche market segment. It is popular for small firms for these reasons: (1) mass production, mass distribution, and mass advertising are not necessary; (2) it can succeed with limited resources and abilities by concentrating efforts; (3) if concentrated marketing is used, it is essential for a firm to do a better job than competitors in several areas (Bhasin, 2016; Horská, 2009).

Opposite of mass marketing, a firm may choose to market one-to-one with a customer, a strategy called micromarketing. A special form of one-to-one marketing, called mass customization, is the targeting of large segments, or traditional mass markets, with highly customized products (Kardes *et al.*, 2014).

A successful segmentation should live up to the following criteria:

- *Effective:* the segments should designate customers whose needs are relatively homogeneous within a segment, but significantly different from those in other segments such that the market is divided into separate segments on the basis of common or shared needs or characteristics that are relevant to the product or service. Segment identification can be made before the fact, by using basic criteria such as demographics, geography, psychographics and the like, or after the fact on the basis of purchase patterns, consumption preferences, and perhaps most currently social media participation.
- *Measurable:* it must be possible to identify customers in the proposed segment, and to understand their characteristics and behaviour patterns. A segment must consist of enough consumers to make targeting it profitable. A segment can be identifiable and stable but not be large enough to be profitable. To measure a segment's size and profitability, marketers use secondary data and consumer surveys.
- *Accessible:* the company must be able to formulate effective marketing programmes for the segments that it identifies. To be targeted, a segment must be accessible, which means that marketers must be able to reach that market segment in an economical way. Even though the market segment under consideration can be identified, measured, and be significant, it may not be accessible, meaning that there may be barriers to entry.
- *Actionable:* the company must have the resources to exploit the opportunities identified through the segmentation scheme. Despite the fact that the market under study may be identifiable, measurable, significant, and accessible, it may not be appropriate for the company analysing it, since the market may not be within the firm's core competency.
- *Profitable:* most importantly, segments must be large enough to be profitable to serve.
- *Stable/significant:* most marketers prefer to target consumer segments that are relatively stable in terms of lifestyles and consumption patterns and avoid fickle segments that are unpredictable. For example, the case of teenagers, they are a size-able and easily identifiable market segment, eager to buy, able to spend, and easily reached (Jobber and Fahy, 2006; Paluchová, 2012; Richterová *et al.*, 2015; Samli, 2013; Schiffman and Kanuk *et al.*, 2010; Světlík, 2003; Vysekalová *et al.*, 2011).

11.3 Bases of segmentation

Figure 11.2 depicts a four-way classification of the characteristics used to segment the buyers of consumer goods. The grouping stems from dividing consumer's characteristics along two

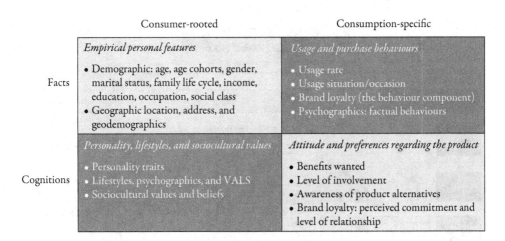

Figure 11.2. Bases of segmentation (adapted from Schiffman and Kanuk, 2010).

criteria: (1) facts, which can be determined from direct questioning and categorized by a single objective measure, vs cognitions, which are abstract, can be determined only through more complex questioning, and where most of the constructs measured have no single, universal definitions; and (2) consumer-rooted features steaming from the consumer's physical, social, and psychological characteristics vs consumption-specific usage-behaviours (Kardes *et al.*, 2014; Schiffman and Kanuk *et al.*, 2010).

Consumer-rooted segmentation bases includes two types of personal attributes: facts, that are evidence-based and can be readily determined and categorized along, and cognitions, which can mostly be determined through indirect, psychological tests and classified into subjective categories, depending on the researcher.

Consumption-specific segmentation bases includes two types of consumption-specific bases for segmentation: facts about actual consumption behaviour and cognitions consumers have about specific products and services in the form of attitude, preferences, and the like (Schiffman and Kanuk, 2010; Solomon, 2011; Vysekalová, 2011).

11.4 Segmenting consumer markets

There are four basic types of consumer segments distinguished by marketers, depending on the major criterion used for segmentation. Figure 11.4 illustrates these four basic types.

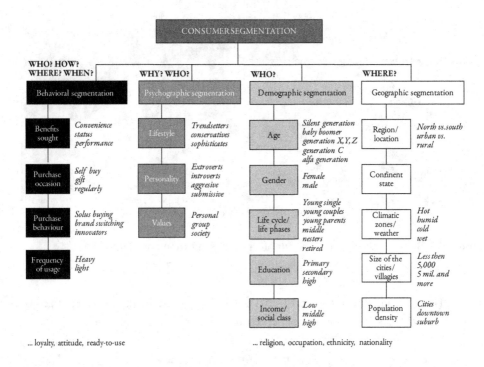

Figure 11.3. Different criteria for segmenting consumer markets.

11.4.1 Behavioural segmentation

In this approach, consumers are divided into groups according to uses or responses to a product or service. Occasion segmentation groups consumers according to special occasions of product use and regularity of use (Rao, 2011). Consumers can be grouped based on their preference for a particular product attribute or benefit, usage occasion, user status, rate of product usage, and loyalty status, buyer readiness stage and attitude (Batra and Kazmi, 2009; Kardes *et al.*, 2014). Unilever actually had to reduce the size of its magnum chocolate-covered ice cream bars sold in China because consumers there resisted the bigger bar, possibly because they were more health-conscious than Europeans. Another example of differences in behavioural habits is seen by tea ceremony across the nations, here are some examples:

▸ *Japan:* 'hot water for tea' and usually refers to either a single ceremony or a ritual. *Cha-ji* or *Chakai* means 'tea meeting' and refers to the full tea ceremony with *kaiseki* (a light meal), *usucha* (thin tea) and *koicha* (thick tea) lasting for approximately four hours. The ceremony of tea is known here as *matcha*, which consists of serving green tea to a small group of people in one of the popular teahouses (consumption 0. 99 kg/annual per capita).

▶ *Argentina:* is associated more with a beverage made with the yerbamate plant, an evergreen in the holly family. The tea is traditionally served warm in a gourd and drunk through a metal straw called *bomba* or *bombilla* (0. 10 kg/ annual per capita).

▶ *Russia:* prefer their tea strong and sweet, and sometimes, served with mint or lemon, or sweetened with fruit jam (1.21 kg/annual per capita).

▶ *China: Gongfu* tea presentation is an important part of everyday life, of social events, business transactions, most favourite is green tea. The best time to drink is in between meals.. Tea drinking is for refreshment and tonic effect. (0. 82 kg/ annual per capita).

▶ *Britain:* the tradition of afternoon tea. Whether it is a short break for a cup of tea and a small cookie, or a 3 course event of cakes, scones with jam and Devonshire cream, sandwiches and other treats, afternoon tea is a true English tradition (2.74 kg/ annual per capita).

▶ *Taiwan:* is similar to the Chinese *Gongfu* presentation, but called the *Wu-Wo* ceremony, where participants bring their own tea and equipage to a scheduled event, take their place in a large circle, and then silently prepare their own tea (1. 85 kg/ annual per capita).

▶ *India:* is called masala chai or spiced tea, where black tea is boiled into a strong infusion and mixed with hand ground spices. India produces and consumes more tea than any other country in the world. *Chai* is the national drink in India and it is served literally on every street corner (0.7 kg/ annual per capita).

▶ *France:* will usually indulge in exquisite pastries, like tarts and *petit fours* with their tea, a tradition that has carried over to many afternoon teas served outside of the country.

▶ *Arabic countries:* Egypt (0.93 kg annual/capita), Jordan (0.79 annual/capita), UAE (1.89 kg annual/capita), Turkey (7.54 kg annual/capita) – black tea, Tunisia (0.92 kg annual/capita), Morocco (4.34 kg annual/capita) – the mint tea is served to guests three times with each glass meaning a different thing life, love and death (Garcia, 2014; ItoEn, 2016; Tea Market Report, 2015; The culture of the world, 2016; consumption data are from 2015).

11.4.2 Psychographic segmentation

It appears that persons with different personality traits hold different beliefs about what is the right choice. Psychographic segmentation is closed related to lifestyle, though the two terms psychographics and lifestyle are not identical, they are close and complementary. Psychographics involves mapping or profiling consumers based on their psychological attributes and evaluating their activities, interests and opinions, and relating these to their purchasing and consuming patterns. Psychographic segmentation refers to the inner or intrinsic qualities of the individual and provides the bases for the repositioning or redesigning of products in the market (Apruebo, 2005). Lifestyle is used to describe the set of general goals a person sets for himself and also the way s/he tries to achieve these goals (Raju and Xardel, 2009). Consumer lifestyle refers to patterns of behaviour reflecting how a person, or more likely a family, chooses to spend their time and money (Wright, 2006). Lifestyle and psychographics are segmentation forms that more closely reflect the way people live and why they buy (Reynolds and Olson, 2001; Schiffman *et al.*, 2014). When lifestyle bridges both cross-nationally relatively stable values and cross-nationally variable product-specific cognitions, we would expect it to contain both nation-specific and cross-national components. To the extent it contains cross-national components; these may give rise to cross-national segments. When analysing consumer lifestyle with regard to food, the following elements have been distinguished: ways of shopping, cooking methods, quality aspects, consumption situations, and purchasing motives (Grunert *et al.*, 2001).

Psychographic segmentation has proven to be a valuable marketing tool that helps identify promising consumer segments that are likely to be responsive to specific marketing messages. The psychographic profile of a consumer segment can be thought of as a composite of consumers' measured activities, interests, and opinions (often referred to as AIOs) (Schiffman *et al.*, 2014). One well-known psychographic segmentation system is 'the values and lifestyle system' (VALS). The VALS system (e.g. Solomon, 2011; pp. 266) distinguishes the following segments:
- *Innovators:* the top VALS group, are successful consumers with many resources, is concerned with social issues and is open to change;
- *Thinkers:* are satisfied, reflective, and comfortable;
- *Achievers:* are career oriented and prefer predictability to risk or self-discovery;
- *Experiencers:* are impulsive, young, and enjoy offbeat or risky experiences;
- *Believers:* have strong principles and favour proven brands;
- *Strivers:* are similar to achievers but have fewer resources, they are concerned about the approval of others;
- *Makers:* are action oriented and tend to focus their energies on self-sufficiency;
- *Strugglers:* are at the bottom of the economic ladder, they are most concerned with meeting the needs of the moment and have limited ability to acquire anything beyond the basic goods needed for survival.

Several alternative systems for psychographic segmentation have been proposed. The list of values (LOV) classifies consumers according to their endorsement of the values self-fulfilment, sense of belonging, security, self-respect, warm relationships with others, fun & enjoyment in life/excitement, being well respected, sense of accomplishment (Kahle, 2000). It has greater predictive utility than does VALS regarding consumer behaviour trends. The LOV has been employed in a wide variety of consumer research settings in a variety of nations. The 'Schwartz Value Survey' (SVS) is often used as the method of choice, because it integrates relevant predecessor systems and alternative systems across cultures (Schwartz, 1994). A Version of the SVS is the means by which the European Commission and the European Science Foundation conduct a biennial European Social

Survey, a Europe-wide analysis of prevalent values. This enables the systematic study of relationships between human values and constructs such as self-reported behaviour, which can be related to consumer engagement in, for example, boycott action. SVS consists of two categories/dimensions (self-enhancement vs. self-transcendence, openness to change vs conservation), ten basic values (power, achievement, hedonism, stimulation, self-direction, universalism, benevolence, tradition, conformity, security) and 58 value items, which are circularly arranged in a value-circumplex (Adugu, 2016; Reisyan, 2015).

Identifying which psychographic segments are most prevalent in certain markets helps marketers plan and promote more effectively. A psychographic study may refine the picture of segment characteristics to give a more elaborate lifestyle profile of the consumers in the firm's target market. For example, Campbell Soup Company (CSC) is a global food company and the world's largest producer of soup with a 60% market share in the 3.59€ billion soup market. It sells the soups *Cajun gumbo* in Louisiana and Mississippi, cheese spice soup *Nacho* in Texas and California, *Duck gizzard* soup for Chinese consumers, *Crema de chile poblano* soup for Mexicans and *Flaki* (a peppery tripe soup) for Poles. England is the most important foreign market for CSC. The English, accustomed to the ready-to-eat Heinz soups but unfamiliar with the concept of condensed soups, were unable to justify the cost of the smaller Campbell's soup can compared to the larger Heinz. Failure to adapt the taste of their soups to local palates was another problem. The taste of established local varieties of tomato soup so differed from Campbell's that it was not until Campbell made significant changes in its flavours that sales picked up. As a result, Campbell's now creates new products to appeal to distinctly regional tastes (Kotler and Armstrong, 2004; Mueller, 2006).

11.4.3 Demographic segmentation

Demographic segmentation refers to the statistical description of a population based on objective criteria (includes age, gender, marital status, income, occupation, education, etc.) (Apruebo, 2005). Demographic segments and most other unidimensional segmentation schemes provide only a piece of the segmentation puzzle (Reynolds and Olson, 2001). Demography refers to the identifiable and measurable statistics of a population. It helps to locate a target market. Demographic information is often the most accessible and cost-effective way of identifying a target market (Schiffman *et al.*, 2014). A clear advantage of this approach over others is that there are vast amounts of secondary data available that will enable marketers to divide a market according to demographic variables. The CACI Market Analysis Group developed ACORN (a classification of residential neighbourhoods), a widely used source for geographic segmentation that uses 40 variables to identity for example what consumers within a specific neighbourhood earn and buy, such as thriving, expanding, rising, settling, aspiring, striving (Market Segmentation, 2016). For example, in terms of purchasing power of the population, a European average is: (1) Liechtenstein (418); (2) Switzerland (283); (3) Norway (233); (4) Luxembourg (220); (5) Denmark (168); (23) Slovakia (57.5); (28) Czech Republic (53.4); (28) Poland (47.1); (31) Hungary (37.7). Now, in Europe 25% of the population is already aged 65+. Many nations do not collect income data; like Great Britain, Japan, Spain, France and Italy. Ireland acknowledges only three marital statuses: single, married, and widowed, while Latin American nations and Sweden count their unmarried cohabitants (Kurtz, 2013).

Sociocultural segmentation is a mix of demographic and psychographic segmentation and refers to subdivisions of consumer markets in terms of variables like family, social class, cultural values, sub-cultures, and cross-cultural affiliation. Sociocultural segmentation includes: (1) culture; (2) subculture; (3) religion; (4) race/ethnicity; (5) social class; and (6) family life cycle (Schiffman *et al.*, 2014).

Take a brand like Jack Daniel's (JD), which is huge and appeals to a very wide variety of consumers. JD's target is fairly broad and has over 3 mil. Facebook fans worldwide. When they advertise on Facebook, they may want the message to go to consumer's ages 25 to 35 years old, or may want to go after women or older consumers. If somebody ID's in their profile that they're Hispanic-American, the company can target them that way, too. When they launched JD Tennessee Honey Whiskey in the USA, they had people in Mexico and the UK saying 'What about us?' They got a query from a customer in Germany asking 'Can we taste it?' Nothing is local anymore. The beverages (whiskey, rum, tequila, vodka, and cognac) have been increasingly aimed at younger consumers, aged between 21 and 35 in the USA. Male and female segments can be targeted by appealing to their preferences for taste, quality, being trendy, and tradition in whisky. Promotion for both is: traditional/ humorous, men: traditional/ warmth, women: trendy/ warmth. Japanese consumers use JD as a dinner beverage. A party of four or five consumers in a restaurant will order and drink a bottle of JP with their meal. Australian consumers mostly consume distilled spirits in their homes. Australians, Germans or Austrians (men and women as well as) prefer to mix JD with soft drinks, mostly with coke and call this mixture 'Jack and Cola'. It is also sold in bottles in retails. British distilled spirit consumers also like mixed drinks, but usually partake in bars and restaurants. In China and India consumers more often choose counterfeit or 'knock-offs' to save money. Chinese consumers enjoy their JD mixed with green tea (E-marketer, 2012; Zikmund and Babin, 2013).z

11.4.4 Geographic segmentation

This is also called marketing by location. The assumption is that consumers who stay in the same location share similar needs and wants (Apruebo, 2005). In geographic segmentation the market is divided by location, nations, states, regions, areas of certain climatic conditions, urban and rural divide (Batra and Kazmi, 2009). The assumption behind this strategy is that people who live in the same area share some similar needs and wants, and that these needs and wants differ from those of people living in other areas (Schiffman *et al.*, 2014). It refers to analytical techniques that combine data on consumer expenditure and other socioeconomic factors with geographic information about the areas in which people live, in order to identify consumers who share common consumption patterns (Solomon, 2011). Geographic segmentation can be a useful strategy to segment markets because it provides a quick overview of differences and similarities between consumers according to geographical unit, can identify cultural differences between geographical units, takes into consideration climatic differences between geographical units, and recognises language differences between geographical units (Market Segmentation, 2016). The company Nielsen Claritas created an approach of geographic cluster called PRIZM (Potential Rating Index by Zip Markets). Examples of PRIZM clusters are young digerats, beltway boomers, the cosmopolitans, old milltowns (Kotler and Keller, 2013).

For example, there are considerable geographical differences regarding eating and timing of eating: In France, restaurants are closing at 10 pm, in Spain they are just opening. In the summer in some parts of Australia a lot of the time is spent outside the house having picnics and barbeques and drinking ice-cold lager, in England this is relatively rare occurrence. The Swedes do most of their family entertaining in the home while in Italy it is not unusual for the whole family, including small children, to eat together in restaurants (Wright, 2006).

Table 11.1 illustrates another segmentation base together with description and examples from international marketing.

Table 11.1. Market segmentation bases and their description (adapted from Apruebo, 2005; Batra and Kazmi, 2009; Keyes, 2010; Schiffman *et al.*, 2014; Vashisht, 2005).

Segmentation base	Description	Example in practice
Use-related segmentation	Consumers are categorized according to rate of usage (heavy, users, medium users, light users, non-users), awareness status and brand loyalty. It contains: (1) usage rate; (2) awareness status; (3) brand loyalty. Categorize consumers in terms of product/ services or brand usage characteristics.	Consumers are loyal to two or three brands. More people now buy from a small set of acceptable brands that are equivalent in their habits. Austrian consumers chooses between Almdudler or Coke, Slovak in Kofola vs Coke, Maltese buy Kinie or Coke, and the Swiss decides between Rivella and Coke. All these drinks are local and for other nations not so delicious at first drinking.
User-situation segmentation	Identifying the occasion or situation explains consumer buying and usage activities. It consists of time, objective, location, and person. It contains: (1) time; (2) objective; (3) location; (4) person. Sometimes the occasion or situation determines what consumers will purchase.	In China, Cadbury Schweppes PLC makes its Cadbury milk chocolate less milky and less sweet compared with that in the UK to suit the low-dairy diet of most Chinese consumers, whereas Kraft adds calcium to the Ritz crackers it sells in China to underscore the Chinese government's efforts to get its citizens to eat more calcium.
Benefit segmentation	Satisfaction of the consumers can be best achieved through the benefits they received or experienced from the products. It requires finding the major benefits people look for in the product class, the people who look for each benefit and the brands that deliver each benefit. Benefit segmentation has the potential to divide markets according to why consumers buy a product. Benefits sought by consumers are more likely to determine purchase behaviour than are descriptive characteristics. Marketers strive to identify the one most important benefit of their product or service that will be most meaningful to consumers.	McDonald's offer busy consumers the benefit of breakfast products and various choices of budget meals that require only seconds of preparation. Benefits: convenience, social acceptance, long lasting economy, and value-for money.

Table 11.1. Continued.

Segmentation base	Description	Example in practice
Hybrid segmentation	Provides marketing or consumer researchers with a more valid, reliable, and meaningful information than can be derived from using a single segmentation variable. It contains demographic/psychographic/ geodemographic combinations of segmentation variables. It is useful in creating consumer profiles, audience profiles and segment mass markets, provide meaningful direction as to which type of promotional appeals.	

11.4.5 Global market segmentation

Since a significant segment of the world's population is either untouched directly by globalization or remains largely excluded from its benefits, it is a deeply divisive and, consequently, vigorously contested process. The unevenness of globalization ensures it is a process experienced far from uniformly across the planet (Surman, 2009). But, on the other side, there are some limitations for world segments, which are regularly changed and they are influenced by many factors depending on each country (for example: migration policy in Europe and its impact on cultural and consumption habits, legislation and legal restrictions which may force companies to change goods, distribution, price, etc., life style and social conditions in each country) (Table 11.2). Traditionally, international marketers segmented world markets based on geopolitical variables (i.e. country segments). This approach has three potential limitations: (1) it is based on country variables and not consumer behaviour patterns; (2) it assumes homogeneity of the country segment; and (3) it overlooks the existence of homogeneous consumer segments that exist across national boundaries. (Omarkulova *et al.*, 2013). Geographic segmentation supposes that nations that are close have various similar habits and behaviour. The world market could be segmented according to economic factors (financial incomes or economics developing of country). Then the states are divided according to political or legal factors, type and stability of government, perception of foreign companies, monetary regulation and range of bureaucracy (Kotler and Armstrong, 2004). Cultural variable make a big difference if a consumer lives near Paris in

France or in the mountains in Nepal. Chinese shoppers differ from French shoppers, and rural shoppers differ from urban ones.

As an example, in a global world, the consumption of mineral water was evaluated in 11 selected countries that consume the most bottled water in 2015:

1. Mexico (bottled water has a lot of varieties, one of the more popular is a flavoured one).
2. Thailand (you can find vending machines selling different types of bottled water).
3. Italy (176 l/ per person, strongest consumer of bottled water in Europe).
4. Belgium (124 l/ per person, Belgians prefer it over tap water, is considered 'hard').
5. Germany (168 l/ per person, Germans prefer bottled water over tap water and over any other drink).
6. UAE (with most of the Arabian countries, they import bottled water heavily, is one of the largest importers of bottled water around the world).
7. France (144 l/ per person, water consumption increases along with urbanization).
8. USA (one of the largest markets for bottled water in the world).
9. Spain (113 l/ per person, 90% of Spain's population prefer bottled water, specifically mineral water).
10. China (many Chinese families still opt to buy bottled water).
11. Lebanon (one of the largest producers and consumers of bottled water in the entire world).

European consumer prefer 44% bottled water, 40% lemonade, 9% juices, and 7% syrups (Caránek, 2014, 2015; Mikulcová, 2016; Roy, 2015).

Global consumer segmentation must cross national lines. First, there could be a global segmentation that is not quite realistic and there cannot be national segmentation, let alone a global one. Global brands or producers have advantages regarding the lower need of adaptation approaches in their business activities (Sprite, Microsoft, Hotpoint, etc.). Second, crossing national lines can be interpreted as markets being homogeneous domestically and segmented internationally (multinational companies, as Mondelez or Nestlé, which produce in each countries but export over the world, through the buying of local brands) (Hassan and Samli, 1994; Paluchová, 2014a). Table 11.3 in general advocates that unlike the conventional wisdom, which emphasized nation-markets and clustered them to create international segments, it is suggested that international segmentation should be based on grouping of micro-segments both within a country or among countries or on emphasizing uniqueness of certain micro-segments both domestic and international. Three critical points for international marketers: (1) segmentation is a prerequisite for success in international marketing; companies cannot be all things for all people and choose of segmentation

Table 11.2. Global segmentation (adapted from Surman, 2009).

Segment/name	Segment/description
World segment	Low price, high quality, part of a world homogenous market
Specialty, product, and market segment	Products adapted to local market, different segments across different markets, product modified from country to country
Country segments	Individual countries represent separate segments
Country groupings or clusters	Identification of country groupings with similar demographic, cultural, and buyer behaviour similarities
Regional segments	Identification of regions with similar characteristics for economics of scale
Cultural segments	Identification of similar cultural values and attributes across country boundaries
Strategically equivalent segments	Segmentation to respond to a specific marketing mix
Pro-trade segments	Segmentation on the basis of attitudes toward imports in developed and developing countries
Two-stage segments	Segment by environmental indicators, and further segment by buyer behaviour indicators
Attitude clusters	Similar consumer attitudes for specific products across countries

criterias; (2) an unconventional approach to segmentation must be utilized; (3) within segments use the same marketing strategy, between segments there is a need for different marketing strategies and their selection (Machková, 2009; Samli, 2013).

For example, since 2007, Finland, Sweden, The Netherlands, Switzerland, Greece (Europe), USA (North America), Kazakhstan Kyrgyzstan, Pakistan (Asia), Australia (Australia), Argentina, Uruguay (South America), and Sudan (Africa) are the world countries with the highest consumption of drinking milk. Figure 11.4 illustrates cross-national segments of drinking milk, distinguished according to consumer habits, preference and customs.

Usunier (2000) defined four basic zones for segmenting European countries. But nowadays, the European Union has new members and the situation was changed, because the Central or Balkan countries began to be important for food producers. Usunier's four zones consist of Scandinavian and Mediterranean countries, the third zones includes German speaking countries and the fourth one is countries of Benelux including Great Britain and Ireland.

Table 11.3. Conventional and unconventional approaches in segmentation (adapted from Hassan and Samli, 1994).

Conventional approach in segmentation	Unconventional approach in segmentation
It assumes heterogeneity between countries.	It assumes the emergence of niche markets that transcend national boundaries.
It assumes homogeneity within any given country.	It dwells upon the differences that exist within countries.
It is based on macro-criteria of segmentation.	It is based on hybrid or modern criteria of segmentation based on consumer behaviour peculiarities.
Heavy cultural focus only at macro-level.	Emphasizing differences in behaviour and values based consumption at the micro-level.
Methods of segmentation are based on clustering nation-markets.	Methods of segmentation are based on grouping or clustering micro-markets within a country or between countries.
Micro-or within-the-country segments are given secondary priority.	Micro-segments based on consumer behaviour are major points of emphasis.
Country-wide market considerations are primary emphasis.	Country-wide market considerations are not very important.
Major emphasis is on similarities.	Major emphasis is on differences.
Social media are used to determine national macro-characteristics.	Social media are used to determine the key characteristics of local emerging niches.

Figure 11.5 shows a large-scale analysis of food cultures in European countries. The French/ Swiss/Wallonian and Italian cultures characterized by the importance of sensory pleasure and high consumption of red wine; the Germanic cluster of countries exhibit a high degree of health consciousness; the Portuguese and Greek food cultures show relatively traditional eating patterns with a fascination for new global food; Norwegian and Danish food cultures are unique in their openness to convenience products in future and Danes also for the love of beer; British and Irish by their extraordinary desire for sweet and tea.

Common characteristics for V4 countries (Poland, Slovakia, Bohemia, Hungary) are typically being price-sensitive regarding food, love of foreign brands, less preference and spending money for bio/ fair-trade or FSC (Forest Stewardship Council) products. Eastern European markets are now much more mature. Consumers now consider their choices and became picky. Their lifestyles are also more similar to those in the West.

Scandinavian countries:
Filmjölk is a Swedish cultured dairy food that is mildly sour in flavor and remarkably versatile. *Viili* originally hails from Sweden but is now found in Finland where it is largely considered a national treasure. *Piimä*, like many of the other cultured dairy foods.

Britain, Italy:
popular a 100 % *horse milk*

Asian markets:
a) *coconutmilk* and
b) *soy-bean milk*
Asians do not use milk in the way that Europeans and Americans do.

India, Indonesia :
a) about 50 % of milk is consumed on -farm,
b) cow is Hindu sacred animal,
c) increasing popularization of low fat milk,
d) traditional buttermilk is only really used
e) in India where it is called *chaas* (often flavored with salt, cumin, chilies and
f) other spices),
g) fesh milk is also called *dahdi*.

Middle and Far East:
a) *dry/powdemilk*
b) *camelmilk*
The milk of the Afar camels in Ethiopia. The belief among the Bedouin of the Sinai Peninsula, is that any internal disease can be cured by drinking camel milk.

Middle and Eastern European countries:
a) *Sheep milk:* Slovakia, Poland, Bulgaria, Romania, (called as Kefir (in Caucasus too), Zákvas, Matzoon (in Armenia too), or Žinčica etc.)
b) *Matsoni:* Bulgaria, Georgia, Armenia or Russia is particularly good for breakfast, but it poured over fruit for dessert.

The Southern United States, Scotland:
a) *Bonny Clabber* is a traditional cultured dairy, it was customarily eaten with molasses, cinnamon and nutmeg for breakfast. Milk is mostly sold in 3 l, 5 l and bigger plastic bottles.
b) *Blaand* is a traditional Scottish drink made from fermented whey with quite an alcoholic kick.

Turkey:
Ayran is a national cold yogurt beverage mixed with salt, usually consumed after spice meal.

Figure 11.4. Segments of milk consumption in different regions of the world (image source: https://tinyurl.com/y8mea979).

Russia is the most important consumer market in the region of Eastern Europe. That means that Eastern European consumers are slower to adopt new trends (in particular technology) and need time to accept new situations. An extreme showcase is the Eastern Europeans' love for fences. Russians are the most frequent buyers of local food and very proud of home food goods.

The most important motives in Balkan and Southern countries were found to be: taste, price, health and natural content of the food. Comparison of consumption in all four categories indicates that consumption of fruit is the most extensive; it is followed by consumption of traditional dishes, while products with health claims are consumed least. Fruit is consumed most on a daily basis in Slovenia, and the least in Serbia. Traditional dishes are consumed most in Macedonia and Serbia and less in Slovenia (Focus Balkans Project, 2016). Serbia and Macedonia have a very high percentage of consumers who never buy organic; more interest for these seems to be in Croatia and Slovenia.

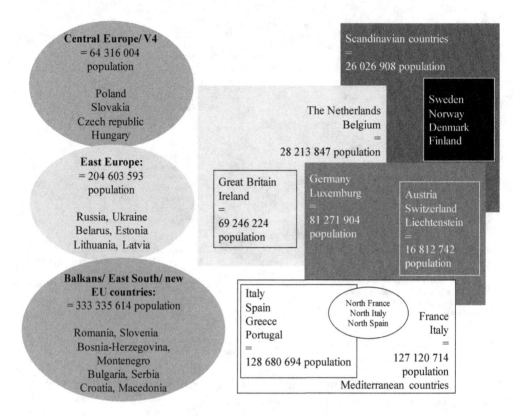

Figure 11.5. Segmentation of the European market (data in boxes adapted from Usunier, 2000; calculation of population adapted from Eurostat, 2015).

Food preferences say a lot about us, and many of our likes and dislikes are learned responses to dishes that people who matter to us value or don't value. A food culture is a pattern of food beverage consumption that reflects the values of a social group. Food companies often find it rough going when they try to standardize their recipes (Berčík *et al.*, 2016). Sometimes these preferences are based on tastes and traditions, but in other cases people in other countries differ in terms of their focus on health (Solomon, 2011). Attitude and behaviour common to consumers across Europe mainly refer to the search for convenience, more variety in foods, better and consistent quality of food products and concern over health (Kaynak, 2010). For the introduction of cholesterol-lowering Benecol margarine, designers created a package that included a mountain scene to communicate cross-culturally that the product's key ingredient is natural. In the USA market the package for the spread featured an English muffin, as opposed to Europe, where rolls, which are more commonly consumed, were depicted. While white colour connotes purity and cleanliness in many regions and cultures, in the Far East it is also associated with mourning. Likewise, red is a positive colour in Denmark as well as many Asian countries, but is represents witchcraft and death in a number of African nations (Mueller, 2006).

11.5 New segments of food consumers

We live in the age of the 'foodie'. International cuisine, cooking TV shows, celebrity chefs, food magazines, food bloggers, organic groceries, cupcakeries and other specialities are trends which have an impact on our consumption. In some developing countries, food is often sold in open markets or in small stores, typically with more locally produced and fewer branded products available. Even in many industrialized countries, supermarkets are less common than they are in the USA or Western European countries. Local neighbourhood stores are popular and impractical to drive to a large supermarket for example in Austria, Bulgaria, Malta, Sicily, south Spain and Greece. Continued interest in home cooking has been driven by cooking shows on TV and by blogging foodies.

▸ *Segment of sustainable and responsible consumers:* Is aware of the information about conditions and consequences of production to developing countries when purchasing goods and also prefers products and companies with better social and environmental performance, i.e. products whose production or consumption has no negative impact on people and the environment in developing countries. Is interested in product's ingredients, disseminates information, can explain why s/he is buying organic products, is influenced by advertising, does not go only for price, but also for quality, prefers fair-trade, FSC and bio products, supports and buys local products, prioritizes voluntary modesty before consumption and trends (Horská *et al.,* 2013; Paluchová, 2014b; Paluchová and Benda Prokeinová, 2013).

▸ *Segment of environmental consumers:* increasingly prefer companies that are environmentally responsible, environmental performance of products becomes an increasingly competitive factor (Paluchová and Benda Prokeinová, 2013). Swedes considered plastics packaging most harmful. Americans marked polystyrene containers and plastic. The Slovak brand Rajec began with a marketing campaign 'Milk needs a glass' or 'From new glass bottle 0.75 l, Rajec tastes better', supported the glass as one of the best pack for drinks. Another brand in Slovakia, Santovka presents two variants of packs: (1) 0.25 l in glass for 0.25€ and (2) 0.25 l in plastic for 0.25€, means the same price for same quantity, but different material. Europe is the biggest producer of glass package in the world. 90% of the Europeans believe that glass keeps taste and nutrition value of food. 86% believe that products in glass have the highest added value.

▸ *Segment of green consumers:* prefer natural products, mostly packaged in green packs. They know eco-brands and environmentally friendly products (Lušňáková and Kleinová, 2012; Paluchová and Benda Prokeinová, 2013). Cars come to be 'green'. In this time, the car companies create these campaigns: Toyota (Clean Air), Daimler Chrysler 'Fresh Air', Kia 'Think before you drive', Peugeot 'Flower' or Honda 'Safe and Environmental'. The brand Coca Cola introduced a few years ago new Coke 'Life' in green pack, with less sugar and added ingredients, the main idea sounds: 'Save the water', 'Rain Water Harvesting', or 'Return back to environment'. In Japan, KitKat (company Nestlé) launched the Green tea flavour snack, also in green pack. Many foods for kids (brand Gerber, Happy Tot, Nutrilon) combine the flavour with a picture, e.g. green beans or spinach illustrated on pack.

▶ *Segment of bio consumers:* exhibits enhanced integrity and enforcement of inspection and certification procedures. Consumer demands with reference to quality and safety of products are envisaged as being met, minimizing the risk of organic food scandals. Organic consumers: both 'regular' and 'occasional' appear to want stronger product, means consumer relationships. Main market substitutes for organic food (ethical, fair trade, local and slow) (Jensen *et al.,* 2011; Vietoris *et al.,* 2016). In the organic market, however, only few national logos (e.g. The Swiss Bud, the Danish logo), few private certification labels (e.g. Demeter) and few private brands (e.g. Rapunzel) currently provide enough equity to foster loyalty. North America and Europe generate the main sale of organic products. These two regions have about one third of global ecological land and still make up more than 90% sales of organic food. Europe is the second largest organic market, worth about 31.4 billion Euro. Germany is the country with the second largest market for organic products in the world, with a value of about 9.4 billion Euro (bio consumption 86€ per capita). The French market (bio consumption 61€ per capita) is the second largest in Europe, followed by the UK (bio consumption 58€ per capita) and Italian (bio consumption 55€ per capita) markets. Other important markets for organic products are Switzerland (bio consumption 189€ per capita), Austria (bio consumption 127€ per capita), Sweden (bio consumption 95€ per capita), Denmark (bio consumption 159€ per capita) and the Netherlands (bio consumption 60€ per capita). The largest increase in the number of bio transactions is in Germany, where every year is opened more than 50 organic supermarkets (data of bio consumption are in 2014, adapted from European Commission (EC, 2016), Euromonitor, BioFach 2014).

▶ *Segment of functional consumers:* Priority to focus on healthy and beneficial food. There is no doubt that in case of the functional food (Figure 11.6) the added value is represented by the added functions and positive health effects. There is a challenge in relation to the production process to add 'extra function or additional purpose' to the existing food products. (Horská and Sparke, 2007).

▶ *Health and well-being consumers:* One of the biggest demands consumers are asking of food manufacturers is for more 'good for you' products. This includes fresher ingredients, organic foods, local products and health positive foods (Wiley, 2016). Promoting healthy diets and lifestyles to reduce the global burden of no communicable diseases requires

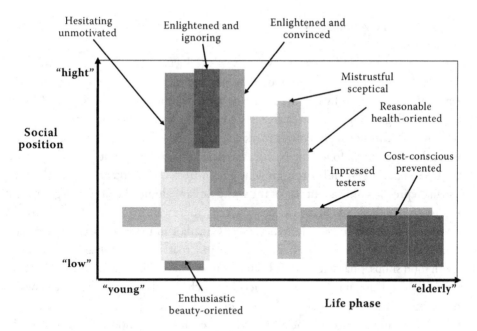

Figure 11.6. Mapping of functional-food-related consumer groups (Horská and Sparke, 2007).

a multispectral approach involving the various relevant sectors in societies towards regional consumption. They buy in open markets or directly on farms. Many of them read the packages and prefer country-of-origin and nutritional information. Many of them have own garden or farm.

▸ *Segment of busy life, taste over health consumers:* within this segment don't spend much time in the kitchen. Many brands reflect to this group and produce a lot of products in cans, frozen, ready-to-eat, etc. Top ten canned food consumer countries in kg per capita in the last five years are: (1) Sweden (33.4); (2) UK (23.5); (3) Portugal (23.2); (4) France (22.3); (5) Belgium (21.4); (6) USA (21.2); (7) New Zealand (16.5); (8) Norway (16.4); (9) Czech Republic (16.2); (10) Australia (16.0). On the other side, Taiwan is currently world's top canned food exporter to Europe, Asia, Australia and North America. Total production of canned food in Russia was in 2014 in 12.2% higher than in 2013. Russia becomes to be canned food consumer (Maps of world, 2016). Formal mealtimes are continuing to decline in popularity and growing numbers of foods and drinks are now considered to be snacks. Quick, healthy foods are tending to replace traditional meal occasions and more snacks are targeted at specific moments of consumption, with different demand influences at different times of day (Global Food Forums, 2015).

An increase in new food delivery methods and drinkable food options are also likely to rise in 2016.

▶ *Segment of diabetics and dieters:* to avoid sweet products and prefer or they have to eat gluten free food. The trend is to stop consume a gluten, because of many promoted negative effects to healthy but the reality is, that the number of consumer with celiac disease is still growing up. Gluten free is the fastest growing food intolerance category in the world. The global gluten-free market is projected to reach US$ 6.2 billion (5. 47 billion Euro) by 2018, following North America contributing about 59% of the share. The USA is the largest and fastest-growing gluten-free (GF) market globally. Although Canada represents only about 4% of the world dollar volume, the market is growing at parallel rates to the USA. Between 2007 and 2015, there were in Canada 2,344, and USA 10,482 new food products and 197 beverage (Canada), and 1,645 (USA) new beverage products with GF claims. The GF products are in Croatia sold mainly in larger cities in health food shops, pharmacies and also some supermarkets of following chains as G.M. Pharma, Billa, Konzum. The GF products in Czech Republic can be bought in health food shops Zdrava vyziva, Natural, Racio, Country life and supermarkets Delvita, Tesco, Albert. They do not have any local brands but they import Mantler flour from Austria. In Denmark, GF products could be bought in Health-food-shops, e-shops and in a few supermarkets. Gluten-free is not also wheat-free in Denmark or Finland. Burgers at McDonald's and pizza at nationwide pizza chain Rosso offer in Finland gluten-free meal. In Germany, GF products are sold specially in shops called 'Reformhaus'. In Italy, the GF products are on sale mainly in pharmacies located in big towns. The Italian Coeliac Society has selected some restaurants and fast food (not belonging to a chain as McDonald's). (Agriculture and Agri-Food Canada, 2016; Coeliac Youth of Europe, 2016).

▶ *Segment of sensory preference consumers:* a profitable way to understand these inter-individual differences is to accept the fact that people differ, and then look for ways to standardize the way they discover and use these differences. (Moskowitz *et al.*, 2012).

▶ *Segment of animal-friendly consumer:* or called vegetarians, vegans or raw-food consumers. They strictly refuse the meat and meat products, vegans also milk and other products made from animal. Nowadays, it is also life style but on the other side, the reason is to protect their healthy or animal life. Increased awareness about the health benefits of seafood has steadily increased its consumption and is often viewed as a healthier option to other protein foods. Aquaculture is stepping in to assist with feeding the growing global population which requires a source of sustainable protein. The aquaculture industry continues to grow due to its viability as a solution to maintaining seafood production in the face of declining ocean seafood stocks (Wiley, 2016).

▶ *Private label powers on:* Even through the worst of the economic recession is over, private label is still gaining market share in terms of new product launches in Europe, North

America and Australasia. Store brands are here to stay and are found in all product segments. Discounters Aldi and Lidl are by consumers no longer solely seen as budget stores, but are accepted by the general public and considered to have good quality products (Global Food Forums, 2015).

References

Adugu, E., 2016. Correlates of political consumption in Africa. In: Gbadamosi, A. (ed.) Consumerism and buying behaviour in developing nations. IGI Global, London, UK, pp. 452-474.

Agriculture and Agri-Food Canada, 2016. 'Gluten free' claims in the marketplace. Available at: http://tinyurl.com/zmhtqrs.

Apruebo, R.A., 2005. Applied consumer psychology. Rex Book Store, Manila, Philippines, 213 pp.

Batra, S. and Kazmi, S.H.H., 2009. Consumer behavior. Text and cases, 2nd edition. Excel Books, New Delhi, India, 544 pp.

Berčík, J., Horská, E., Wang, R.W.Y. and Chen, Y.C., 2016. The impact of parameters of store illumination on food shopper response. Appetite 106: 101-109.

Bhasin, H., 2016. Concentrated marketing. Availabe at: http://tinyurl.com/ycp7h6n5.

Boone, L.E. and Kurtz, D.L., 2013. Contemporary marketing, 16th edition. Cengage Learning, Boston, MA, USA, 784 pp.

Caránek, J., 2014. Európania minerálky milujú, Slovensko zaostáva. Obchod Journal 4: 26-27.

Caránek, J., 2015. Trh balených vôd ožil. Obchod Journal 6: 18.

Coeliac Youth of Europe, 2016. Working group know-how 'How to get gluten-free food in Europe'. Available at: http://tinyurl.com/ycx2ng37.

E-marketer, 2012. Jack Daniel's parent targets a varied customer base. Available at: http://tinyurl.com/yb2kctsp.

European Commission (EC), 2016. Organic farming. Available at: http://tinyurl.com/y9qsra9z.

Eurostat, 2015. Population. Available at: http://tinyurl.com/yb6ot9e2.

Focus Balkans Project, 2016. Characteristics of food consumers in western Balkans countries, results of a quantitative survey in six WBCs. Available at: http://tinyurl.com/y7pctecs.

Garcia, M.L., 2014. 10 tea traditions around the world. Available at: http://tinyurl.com/y7mspmnx.

Global Food Forums, 2016. Food trends. Available at: http://tinyurl.com/y96hke7m.

Grunert, K.G., Brunsø, K., Bredahl, L. and Bech, A.C., 2001. Food-related lifestyle: a segmentation approach to European food consumers. In: Frewer, L.J. Risvik, E. and Schifferstein, H. (eds.) Food, people and society. A European perspective of consumers' food choices. Springer-Verlag, Berlin, Germany, pp. 211-230.

Hassan, S.S. and Samli, A.C., 1994. New frontiers of intermarket segmentation. In: Hassan, S.S. and Blackwell, R.D. (eds.) Global marketing. The Dryden Press, New York, NY, USA, pp. 76-100.

Horská, E. and Sparke, K., 2007. Marketing attitudes towards the functional food and implications for market segmentation. Agricultural Economics – Czech 53: 349-353.

Horská, E., 2009. Európsky spotrebiteľ a spotrebiteľské správanie. SUA, Nitra, Slovakia, 219 pp.

Horská, E., Paluchová, J., Prokeinová, R. and Moiseva, O.A., 2011. Vnímanie imidžu krajiny pôvodu potravinárskych produktov a aspekty ich kvality vo vybraných európskych krajinách. SUA, Nitra, Slovakia, 158 pp.

Horská, E., Yespolov, I.T. and Gálová, J., 2013. Sustainability: communicating, reporting and managing change. In: Horská, E. and Yespolov, I.T. (ed.) Sustainability in business and society: global challenges – local solutions. Wydawnictwo Episteme, Kraków, Poland, pp. 153-166.

ItoEn, 2016. All things tea. Available at: http://tinyurl.com/hnleh8h.

Jensen, K.O.D., Denver, S. and Zanoli, R., 2011. Actual and potential development of consumer demand on the organic food market in Europe. NJAS – Wageningen Journal of Life Science 3: 79-84.

Jobber, D. and Fahy, J., 2006. Foundation of marketing, 2nd edition. McGraw-Hill Education, Berkshire, UK, 376 pp.

Kardes, F.R., Cronley, M.L. and Cline, T.W., 2014. Consumer behavior, 2nd edition. South-Western College Pub, Boston, MA, USA, 576 pp.

Kaynak, E. and Kahle, L.R., 2000. Cross-national consumer psychographics. Routledge, New York, NY, USA, 137 pp.

Kaynak, E., 2010. Cross-national and cross-cultural issues in food marketing. Routledge, New York, NY, USA, 146 pp.

Keyes, J., 2010. Marketing IT products and services. CRC Press, Boca Raton, FL, USA, 336 pp.

Kotler, P. and Armstrong, G., 2004. Marketing. Grada Publishing, Prague, Czech Republic, 855 pp.

Kotler, P. and Keller, K.L., 2013. Marketing management. Grada Publishing, Prague, Czech Republic, 814 pp.

Kubicová, Ľ., 2013. Strategický marketing. SUA, Nitra, Slovakia, 169 pp.

Lušňáková, Z. and Kleinová, K., 2012. The place of corporate social responsibility and its activities in the retail firm management. Zeszyty Naukowe Szkoły Głównej Gospodarstwa Wiejskiego w Warszawie Journal 56: 53-61.

Machková, H., 2009. Mezinárodní marketing. Grada Publishing, Prague, Czech Republic, 196 pp.

Maps of World, 2016. Top ten canned food consumer countries. Available at: http://tinyurl.com/ycrcek64.

Market Segmentation, 2016. Segmentation strategies. Available at: http://tinyurl.com/yb5so2vq.

Mikulcová, K., 2016. Nealko nám chutí. Obchod Journal 1: 8-9.

Moskowitz, H.R., Beckley, J.H. and Resurreccion, A.V.A., 2012. Sensory and consumer research in food product design and developmnent, 2nd edition. Wiley-Blackwell, New York, NY, USA, 440 pp.

Mueller, B., 2006. International advertising. Theoretical and practical perspectives. Peter Lang Publishing Inc., New York, NY, USA, 368 pp.

Omarkulova, M., Paluchová, J. and Uspanova, M., 2013. International cultural environment. In: Omarkulova, M., Horská, E. and Uspanova, M. (eds.) Selected issues in international marketing. Turan University, Almaty, Kazakhstan, pp. 83-96.

Paluchová, J. and Benda Prokeinová, R., 2013. Udržateľné tendencie v spotrebiteľskom správaní: asociačné pravidlá, udržateľný marketing a zodpovedná spotreba. SUA, Nitra, Slovakia, 112 pp.

Paluchová, J., 2012. The new approach on food quality: an important factor on consumer behaviour and trends in food quality. Zeszyty Naukowe. Polytiki Europejskie, Finanse i Marketing Journal 8: 355-366.

Paluchová, J., 2014a. Selected aspects of doing business in visegrad countries: Slovak Republic. In: Horská, E. (ed.) International marketing. Within and beyond Visegrad borders. Episteme Wydawnistwo, Krakow, Poland, p 94.

Paluchová, J., 2014b. Fair trade – Koncepcia spravodlivého obchodu. In: Svitačová, E., Cenker, M., Lajdová, Z., Mravcová, A., Moravčíková, D., Paluchová, J., Pechočiak, T., Porubčan, P., Rajčániová, M., Rovný, P. and Badalyan, G. (eds.) Globálne rozvojové vzdelávanie pre ekonómov. SUA, Nitra, Slovakia, pp. 133-143.

Consumer trends and new product opportunities in the food sector

Raju, M.S. and Xardel, D., 2009. Consumer behavior. Concepts, applications and cases. Vikas Publishing House, New Delhi, India, 227 pp.

Rao, K.R.M., 2011. Service marketing, 2nd edition. Dorling Kindersley Ltd., Noida, India, 206 pp.

Reisyan, G.D., 2015. Neuro-organizational culture: a new approach to understanding human behaviour and interaction in the workplace. Springer, Cologne, Germany, 317 pp.

Reynolds, T.J. and Olson, J.C., 2001. Understanding consumer decision making. The means-end approach to marketing and advertising strategy. Lawrence Erlbaum Associates Publishers, Mahwah, NJ, USA, 466 pp.

Richterová, K., Klepochová, D., Kopaničová, J. and Žák, Š., 2015. Spotrebiteľské správanie. Sprint, Bratislava, Slovakia, 404 pp.

Roy, V., 2015. 11 countries that consume the most bottled water. Available at: http://tinyurl.com/ycl5jcwq.

Samli, A.C., 2013. International consumer behaviour in the 21st century. Impact on marketing strategy development. Springer, New York, NY, USA, 192 pp.

Schiffman, L., O'Cass, A., Paladino, A. and Carlson, J., 2014. Consumer behaviour, 6th edition. Pearson, Melbourne, Australia, 476 pp.

Schiffman, L.G. and Kanuk, L.L., 2010. Consumer behaviour, 10th edition. Pearson Education Inc., Upper Saddle River, NJ, USA, 592 pp.

Schwartz, S.H., 1994. Are there universal aspects in the structure and contents of human values? Journal of Social Issues 50: 19-45.

Solomon, M.R., 2011. Consumer behavior. Buying, having, and being, 9th edition. Pearson Education Inc., New York, NY, USA, 680 pp.

Surman, E., 2009. The global consumer. In: Parsons, E. and Maclaran, P. (eds.) Contemporary issues in marketing and consumer behaviour. Routledge, Oxford, UK, pp. 197-211.

Světlík, J., 2003. Marketing pro evropský trh. Grada Publishing, Prague, Czech Republic, 272 pp.

Tea Market Report, 2015. Global tea statistics. Available at: http://tinyurl.com/ycqg8054.

The Culture of the World, 2016. Tea cultures. Available at: http://tinyurl.com/yapootql.

Ürgeová, J. and Horská, E., 2009. Consumer preferences and retail market segmentation and positioning. International Conference on Applied Business Research. September 21-25, 2009. Valletta, Malta, 234 pp.

Usunier, J.C., 2000. Marketing across culture. Pearson Education Ltd., New York, NY, USA, 648 pp.

Vashisht, K., 2005. A practical approach to marketing management. Atlantic Publishers and Distributors, New Delhi, India, 212 pp.

Vietoris, V., Kozelová, D., Mellen, M., Chreneková, M., Potclan, J.E., Fikselová, M., Kopkáš, P. and Horská, E., 2016. Analysis of consumer preferences at organic food purchase in Romania. Polish Journal of Food and Nutrition Sciences 66: 139-146.

Vysekalová, J., Juříková, M., Kotyzová, P. and Jurášková, O., 2011. Chování zákazníka. Jak odkrýt tajemství 'černé skříňky'. Grada Publishing, Prague, Czech Republic, 356 pp.

Wiley, 2016. Top trends consuming the food industry in 2016. Available at: http://tinyurl.com/yb96ex7g.

Wright, R., 2006. Consumer behaviour. Thomson Learning, London, UK, 512 pp.

Zikmund, W.G. and Babin, B.J., 2013. Essentials of marketing research, 5th edition. South-Western Cengage Learning, Chicago, IL, USA, 480 pp.

12. Global food product development

E. Horská[1]*, A. Krasnodębski[2] and R. Matysik-Pejas[2]

[1]*Slovak University of Agriculture in Nitra, Tr. A. Hlinku 2, 949 76 Nitra, Slovak Republic;* [2]*Agricultural University in Kraków, Al. Mickiewicza 21, 31-120 Kraków, Poland; elena.horska@gmail.com*

Abstract

Globalization is a most complex process occurring simultaneously in the economic and social sphere of the modern world. In the economic sphere the process is perceivable as market mergers and internationalization of production, but also as a necessity of adjusting enterprise marketing activities to these conditions. Globalization processes did not omit food markets and their beginning dates back to the 19th century. One of the basic and most important instruments of competition on the global food market are the product and its brand. In case of food, the culture factor plays an important role in creating products for global markets. Even in culturally similar geographical areas, local preferences and customs associated with the consumption of food products should be considered. Therefore, the product must be accompanied by properly developed strategies, allowing the enterprise to adjust its offer to the conditions on foreign markets. The fundamental issue is the scope of assumed standardization or adaptation of the product to the needs and expectations of its potential consumers, which may be of a universal character or diversified in the perspective of countries and regions. Globalization processes are not without influence on changes of the consumer values system. The globalization of food markets also causes the penetration and assimilation patters of consumption on a transnational scale. Globalization of consumption brings at the same time positive and negative outcomes in consumption.

Keywords: globalization, food, product strategies, consumption

12.1 The essence of food market globalisation

Globalization is a current civilizational megatrend in the world economy. In general terms, it stands for an ongoing process of intensification of economic, political and cultural cross-border relations (Naisbitt, 1997). Globalization is a process in which events taking place in one part of the globe have consequences for persons, enterprises and whole economies in other, often very distant parts of the world (Sztucki, 1998).

The economic side of globalization is the most frequently perceived, manifested as integration of markets, forming international and worldwide economic organizations, connecting of regions, companies and societies from different countries and continents. Economic globalization also causes quantitative and qualitative changes in consumption, at the same time creating new trends illustrating the nature of changes and giving them new directions (Olejniczuk-Merta, 2014).

Economic globalization is a process of removing barriers from market operations, resulting in changes in the way in which markets function (Szymański, 2004). Economic globalization is also defined as a progressive integration process of countries and regional markets into a coherent, global market of commodities, capital and services. The process results in the fusion and merging of markets, internationalization of production, distribution and marketing, and adoption of global operational strategies by enterprises (Liberska, 2002).

The globalization of food markets means a process in which food chains become more extended and more complex. These chains are no longer limited to the area of a specific country. This results in a significant increase in the gap between place of food production and place of its consumption (Friedland, 2003).

Trade in food and agricultural products has been one of the fundamental areas of trade in goods between countries for a long time. Imports of certain agri-food products have been a way to supplement domestic production or widening the assortment in the internal market for centuries. However, most food products were produced and sold on the domestic markets (Dybowski, 2005; Friedland, 2003). The process of globalisation of food markets started in the 19th century, but accelerated after the Second World War (Table 12.1). The opening of the economies of individual countries was the result of the signing of the international agreement which sets the rules of global trade (General Agreement on Tariffs and Trade; GATT). As a result, the liberalization of trade took place, which means the reduction of barriers to the flow of goods between countries. The gradual removal of barriers to trade in goods and capital has influenced the development of transnational corporations, and with them global products and brands of food. The globalization of food markets has also caused

Table 12.1. Factors stimulating globalisation (Sztucki, 1998).

Kinds of factors	
Market factors	creation and development of transnational enterprises
	unification of consumer needs in many countries
	opportunities to use homogenous elements of marketing at a global level
Cost factors	benefits resulting from the scale and range of production and trade
	benefits resulting from acquiring experiences and know-how
	benefits resulting from centralized logistics
	access to cheap sources of supply and labour force
Political factors	liberal trade policy
	international technical, ecological and safety standards
	integration among countries
	policy of mother country supporting investment activities
Competitive factors	competitive hazard on a domestic market or on previous foreign markets
	extending economic activity
	ability to compete with integrated marketing on international markets

the penetration and assimilation patterns of consumption on a transnational scale. It leads to new qualities of links between enterprises and markets (Liberska, 2002).

However, it should be remembered that globalization processes penetrate only some regions of the world, whereas others remain wholly out of their reach (Czerny, 2005). All countries that are unable to meet the requirements associated with creating the appropriate hard and soft infrastructure or are unable to supply their economies with advanced resources or skills crucial for contemporary international economic activity are marginalized (Szymański, 2004).

12.2 Product strategies on global markets

In case of food, the culture factor plays an important role in creating products for global markets. Cultural aspects significantly affect consumer buying behaviour. Even in culturally similar geographical areas, local preferences and customs associated with the consumption of food products should be considered. That is particularly important when a company enters more culturally distant markets. Cultural differentiation or similarity of markets, where companies intend to operate, is essential in the process of market selection and product development strategies (Horská, 2014; Zięba, 2010).

The strategies implemented on the global market are multidimensional. For an enterprise that wants to operate on a global market, determining the strategy requires taking a number of parameters into consideration. There are five such basic parameters (Yip, 2004):

▶ market choice: choice of markets on which the enterprise focuses its attention; the level of the operations developed by the enterprise; share in the market;

▶ products: degree of standardization and diversification of products and services offered by an enterprise on markets in various countries;

▶ localization of the activities: the choice of location, where each activity adding value in the food chain will be located – from research and development work to after-sales service;

▶ marketing: using global or local brands, advertising and other marketing elements in various countries;

▶ competitive moves: treating competitive operations in a given country (region) by an enterprise as a part of its global competitive strategy.

A product is one of the fundamental and most important competitive parameters on the global market. Product strategy is adopted by an enterprise as a coherent conception of creating offers. The strategy is the core of the enterprise's operations in all spheres of its functioning in relation to the markets it wants to serve. The enterprise makes decisions on the kinds of products that will be marketed and, in consequence, also on the other marketing instruments supporting the products within the strategy. These activities must ensure communication of the offer and its accessibility (Sudoł *et al.*, 2000).

The marketing concept of a product is composed of a set of features comprising the functional values of the product, its quality, packaging and brand. The prerequisite for success of the adopted product concept on the international market is a strong relation of these features with the needs and expectations of potential buyers, which are partly universal in nature and partly diversified according to the profile of countries and regions (Horská *et al.*, 2014). Before making a decision about the share of product features that may be of a global nature and those that should also consider local requirements of potential customers, an exhaustive analysis of both the product and market should be conducted (Mazurek-Łopacińska, 2015). Proper shaping of the product strategy is the key issue in this respect and the main problem is the extent of product standardization or product adaptation (Figure 12.1).

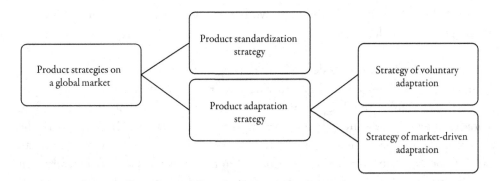

Figure 12.1. Division of product strategies implemented on the international market according to the degree of changes in the product.

Product standardization is the strategy which assumes that a product can be introduced by an enterprise to new foreign markets unchanged compared to the product offered on the domestic market or on other previously serviced foreign markets (Limański and Drabik, 2010). This strategy can also be adopted if it is a possibility to shape consumer preferences through transforming the consumption pattern, possibly based on a positive image of the country of origin of the product (Oczkowska, 2006).

A standardization strategy facilitates maintaining a uniform product and firm image on the international markets serviced by the enterprise, but also contributes to lowering the costs involved in production, distribution and marketing (Wach, 2003). Moreover, there are other arguments in favour of standardization: global competition, advantages of economy of scale, synergistic and transferable experience, easier planning and control, uniformity of consumer preferences, easier communication, planning and control of operations, reduction in supply costs. Moreover, owing to consumer mobility, the strategy may encourage greater consumer loyalty for some products. Accessibility of the same offer in different countries favours this (Hollensen, 2007; Vrontis and Thrassou, 2007).

Because of the benefits that the standardization strategy provides, enterprises are often interested in its implementation on the widest possible scale. However, certain circumstances must arise before it can be implemented. Firstly, a global market segment to which an enterprise will target its offer must exist. The most important aspect is similarity of buyers' needs and behaviours caused by, among others, a similar level of affluence, a similar system of values or even the same climate (Fonfara et al., 2014). Secondly, the strategy must be linked to some expected synergistic effects for the enterprise. The standardization strategy may be applied by enterprises with a high level of internationalization, which have a strong market position (Table 12.2).

Acceptance of global products depends on the degree to which they match the existing consumer trends, but also on the degree to which the products would be creating these trends (Mazurek-Łopacińska, 2003). Consumer willingness to accept global products depends to a large extent also on the strength of their attachment to local culture. The consumers who are strongly embedded in local culture have a lower tendency to accept global products. Culture-bound products are strongly connected to the culture of nations, therefore their standardization is difficult. These are, among others, food products and those, whose role is crucial for expressing the cultural identity of nations. Products of this type require often adaptation to local conditions, because nutritional habits are durable and stabile, so they will not change fast. Nutritional habits are connected to the culture of a given society and a change of nutritional habits may require a change of cultural patterns (Goryńska-Goldman, 2009). Therefore, culture-free products are the most susceptible to standardization, since they have no bonds with the national culture of the target countries. Beverages (e.g. Coca-Cola and Pepsi-Cola), alcohols (e.g. Finlandia, Absolut), tea (e.g. Lipton), coffee (e.g. Nescafe), cigarettes (e.g. Marlboro) or chewing gums (e.g. Wrigley) are among the food products, which gained worldwide success, have a global range and in most countries are sold in a standardized form.

Table 12.2. Environment factors implicating product standardization and adaptation (Wiktor *et al.*, 2008).

Factor	Factors driving product standardization	Factors driving product adaptation
Macro environment		
Economic	Similar level of development	Different level of development
Political and legal	Similar	Different
Technical and technological	Rapid changes	Slow changes
Cultural	Small cultural distance	Big cultural distance
Micro environment		
Character of consumer preferences	Homogenous	Heterogeneous
Consumer purchasing power	Equal	Diverse
Demand for diversified products	Low	High
Competition in sector	Weak	Strong
Range of competition	Global	Local
Enterprise market position	Dominant	Non-dominant

Development of transnational market segments is not a universal trend. Despite the benefits caused by the possibility of manufacturing and offering standard products on the international market, their marketing is in many countries very difficult, sometimes even impossible. The necessity to adapt products results mainly from the fact that the world market remains a set of different local markets that require adjustments of the product offer (Drabik and Limański, 2010). Product standardization is made difficult by the differences that exist between individual markets. They are of a political-legal, social, economic and technological nature. The differences also relate to legal regulations (e.g. safety standards, patent protection, the information on the packaging, standards regarding the contents and form of advertisement, limitations for the application of some media), incomes, language, habits, adopted system of values, technological advancement and market maturity (Fonfara *et al.*, 2014; Wierzbicka, 2012).

Adaptations of some food characteristics when introduced by enterprises on local markets are explained by different needs, preferences and requirements of their consumers. Additionally, socio-cultural differences, differences in lifestyles, religion, or even climatic factors result in that the product must be adapted to the market to find purchasers (Horská *et al.*, 2007; Vrontis and Thrassou, 2007). In such cases, the adjustment of a product to the market, i.e. its adaptation involving product modification, is a necessary measure. The outcome is offering various versions of the same product to different markets (Limański and Drabik, 2010).

Two varieties of the adaptation strategy should be mentioned (Adamczyk and Witek, 2008):
▸ Voluntary adaptation – is largely controlled by the company; its application is dominated by cultural and economic factors, e.g. the income level, education level, consumer preferences in individual countries.
▸ Market-driven adaptation – adjusting the product features to the foreign market requirements, irrespectively of whether the company wants to make these changes or not; the changes in production are most frequently caused by legal regulations and technical conditions (e.g. the regulations regarding the information placed on the packaging of food products).

While marketing Oreo cookies in China, the Kraft Food Company had to adjust their sweetness level to local requirements. Consumers from this country prefer lower sugar content in sweets. Moreover, due to the lower real income of a Chinese family, the company introduced new, smaller packages of the cookies. On the other hand, manufacturers of chewing gum in pellet form destined for the Japanese market pack each pellet in a separate paper to provide the customers with a feeling of safety and sterility of product (Fonfara *et al.*, 2014).

Coca-Cola Concern had to change its name from Diet Coke to Coke Light or to Coca-Cola Light, because in some countries the word 'diet' suggested that the product had weight reducing properties and not only a reduced calories content. Coca-Cola currently uses the 'light' word version on European markets (among others in Poland, Germany and Spain), on South America markets (Argentina, Brazil) and in Mexico.

Not only the product name, but also symbols placed on packaging have some established meaning. A French company exporting cheese from the Pyrenees to German markets put a picture of a shepherd surrounded by sheep on the packaging. It used the same symbols in television advertisements of its product. In France, the image was associated with natural manufacturing processes and traditional values, however German consumers associated the shepherd mostly with dirt. The product was accepted on the market only when the shepherd image was replaced by a mountain countryside (Usunier and Lee, 2005).

Another example is Danone yoghurts, available on the Polish market. Some lines of this product have higher fat contents and are more sweet than the yoghurt offered by the same company on French markets. The reason is Polish consumer preferences, who like a more pronounced taste.

In manufacturing coffee, Kraft prepares a different blend for the British, who prefer drinking coffee with milk, a different blend for the French, who drink black coffee and still another for the Hispanics, who like the taste of chicory (Żak, 2009). The enterprise that decides to operate on international markets and accordingly must choose an appropriate product strategy must consider both the possibilities and limitations of each (Table 12.3).

Both standardization and adaptation strategy must be implemented through a more detailed approach to foreign market support (called EPRG), i.e. ethnocentric, polycentric, regiocentric and geocentric strategy (Figure 12.2).

Ethnocentric strategy assumes that it is possible to transfer the product strategy realized on a domestic market to a foreign market. The assumption results from the enterprise's conviction that its offer is much better than the one accessible on the foreign market and that the way by which consumers evaluate products on the foreign market is similar to the one used on the domestic market. However, these assumptions are quite risky, especially when an enterprise is entering the market that is significantly different culturally and socially from the mother country market. The differences may affect consumer behaviours considerably, and they may not be able to accept the product (Drachal, 2014; Wiktor, 2006).

Table 12.3. Advantages and disadvantages of standardization and adaptation (Fonfara *et al.*, 2014).

	Advantages	Disadvantages
Standardization	• Using similarities in consumer behaviours in various countries • Possibility of cost reduction due to the economies of scale • Simplification of management processes • Opportunity to introduce the same promotional campaign • Merging product image with the company image • Reduction in R&D expenses	• Too serious limitation of the offer • Hazard of a price war on the part of competitors • Ignoring the real needs of the local consumers
Adaptation	• Possibility of supporting niche markets • Greater possibility to diversify the price level • Diversification of products in relation to competitive firms' products	• Higher production costs • Necessity to apply a different promotional strategy

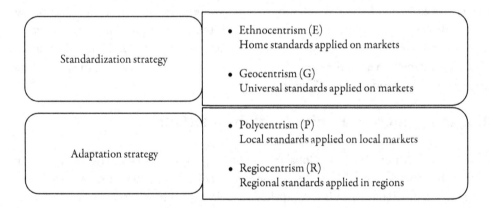

Figure 12.2. EPRG (ethnocentric, polycentric, regiocentric and geocentric strategy) model in standardization and adaptation strategies.

Polycentric strategy is also called multi-local strategy. It assumes a strict adjustment of the enterprise's offer to specific conditions of the markets where it intends to operate. It takes local specificity of level of economic development, social and cultural circumstances, but also other factors of the external environment into consideration. The enterprise must create as many different offers, as many foreign markets it wants to support, because each market is treated differently than the other. It is a quite expensive strategy, but provides big chances of success and gaining competitive advantage (Drachal, 2014).

The regiocentric strategy is a variety of the polycentric strategy. In this case an enterprise's offer targets a group of countries constituting a region for the international operation of the enterprise. The strategy assumes that groups of countries may constitute a common market for the enterprise's activities, because there are more similarities among them than differences, so they are equally susceptible to the nature of stimuli, which the marketing offer contains. This strategy treats all countries in a given region equally. It is an approach providing an opportunity to obtain scale effects, but at the same time it carries the risk of an excessive averaging, which may be too distant from the actual needs of individual markets (Wiktor, 2006).

The geocentric strategy denotes that an enterprise uses a uniform offer for all national markets targeted, irrespectively of their social and economic differences. These markets are treated as a global market and the existing differences are ignored purposefully. The premise for the success of this approach is universalism of the offer and likelihood of consumer acceptance. It means that consumers have similar needs and want to satisfy them in a similar way. The strategy is characteristic for transnational corporations conducting their economic activities on the world market (Wiktor, 2006).

12.3 Innovation strategies on global markets

An enterprise that operates on a global market is not always able to standardize or adapt its offer to the conditions on this market. Sometimes, in order to maintain its previous market position and develop further, it is compelled to seek new solutions.

Innovation strategy is the most difficult aspect of the product strategy on international markets.. Product innovations can include modernized or new products with enriched functional features that better satisfy customer requirements. New food products may be classified according to whether they are a development of an existing line of products, a change in the intended product use, a new form of an existing product, a change of packaging, a change of value added to the product, etc. (Lenart, 2008).

Two types of innovation strategies may be distinguished, i.e. innovation leadership strategy and imitation strategy (Wiktor *et al.*, 2008) (Figure 12.3).

An innovation leadership strategy is an on-going and systematic process of putting new product on the international market. It is used by large enterprises that are market leaders, able to get and maintain their advantage resulting from applied innovations over a longer period of time (Adamczyk and Witek, 2008). The enterprises that implement an innovation leadership strategy may expect benefits concerning the strengthening of their position on the market, improvement of the company's image, possibility to impose some standards on the competition. They may also achieve positive financial results, due to higher prices for their innovative products. An innovation leadership strategy may be realized in many ways, among others through (Grzegorczyk, 2005; Wiktor *et al.*, 2008):

▸ activities aimed at shortening the product life cycle, e.g. through suspending production or sales of older product lines to other companies (the enterprise is not waiting to reach the saturation stage or decline in sales, but sells the license at the final phase of sales growth);
▸ sales of licenses for new products;
▸ forming strategic alliances with other enterprises to work together in order to create new products.

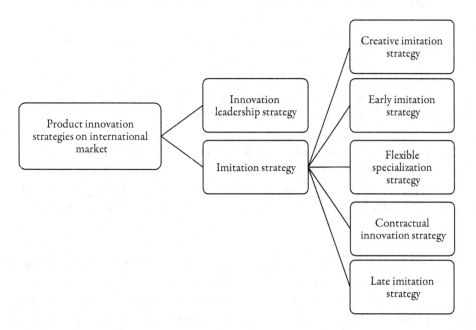

Figure 12.3. Distinction of product innovation strategies on international market according to their degree of innovation originality.

The innovation leadership strategy is rarely used for food products.

The imitation strategies involve that an enterprise responds to the activities of a market leader. They have various forms depending on the enterprise's goals, its market position, resources and potential.

The creative imitation strategy relies on gaining a strong position on a developing new product market. It happens as a result of the innovation diffusion process initiated by the innovator enterprise, i.e. popularization of a new product among consumers on foreign markets. While implementing this strategy, the enterprise uses the experience of the innovator and its new product in order to market improved products or products with alternative characteristics. The strategy requires a well-developed R&D department at the enterprise, considerable outlays on research and other well developed marketing activities (Wiktor *et al.*, 2008).

The early imitation strategy implies that an enterprise supplies new products to the market due to purchases of licenses, patents, know-how or by copying the solutions of other companies as early as possible. Implementation of this strategy is connected with relatively high outlays on acquisition of new technologies and requires a considerable efficiency in starting up the manufacturing of new products (Adamczyk and Witek, 2008).

The enterprise using the flexible specialization strategy modifies the features and properties of products previously offered by the innovator. These activities aim to adjust these features to the needs of specific market segments (Grzegorczyk, 2005; Taranko, 2015). The strategy is applied by medium-sized and small enterprises, because they have relatively small resources and should seek niche markets.

The contractual innovation strategy involves introducing new product innovations commissioned by other enterprises, usually following the patterns and prototypes of the contracting party. The enterprise that implements this strategy indirectly participates in marketing novelties on the foreign market, but not always at its own initiative (Wiktor *et al.*, 2008).

The late imitation strategy involves the gradual introduction of small improvements supported by other measures diversifying the product, but much later than the innovator. The company competes mainly on price and the strategy may be efficient for the market segments which accept the novelties later on, the so-called marauders. This may also be applied on the markets in countries with a lower level of technical development (Grzegorczyk, 2005; Taranko, 2015).

It should be remembered that not all product innovation activities are successful. In the food sector failure rate of new products is between 60 and 80% (Grunert and Valli, 2001). Some of them fail as a result of wrong marketing assumptions (e.g. wrong assessment of potential market, erroneous estimation of promotion and distribution costs, and choice of inappropriate distribution channels, too high or too low a price), technical immaturity of a new product or too long period of innovation implementation in production.

12.4 Global product brand

The brand, regarded as one of the basic marketing tools, is the characteristic feature of a global product. Most brands of food products with a global reach belongs to transnational corporations (TNCs). Every day millions of people around the world buy products of such brands as: Coca-Cola, Fanta, Sprite, Pepsi, Lipton, Lays, Milka, Nescafe, Nestle, Tchibo, Jacobs, Wrigley's, Orbit, Mars, Snickers, M&M, Knor, Gerber, Helmans, Danone, Oreo, McDonald's, KFC, Burger King, Starbucks and many others.

A good, recognizable brand helps to gain and maintain markets and determines the enterprise's value. A brand is a combination of many elements: the name, logo, slogan and packaging developed in order to identify a given product. The source of the brand strength are communication effects (awareness and image) and behavioural effects (behaviours composing the brand loyalty) (Kall *et al.*, 2006).

Brands are equipped with so-called brand capital, composed of the name and symbol, reputation, recognition, functionality and emotions (Badowska, 2014). Global brands carry positive associations formed as a result of their positioning in the buyers' awareness and are viewed as a guarantee of quality, ensuring customer loyalty. Global brands also fulfil very important communication functions as symbols with which persons from various cultural and geographical zones may identify (Patrzałek *et al.*, 2010).

The brand is regarded by many consumers as a crucial element of a product and can increase its value. The development of a brand product is not an easy task and requires long-term marketing investments, designing appropriate promotional actions and advertising campaigns (Kotler *et al.*, 2002).

Brands create a promise of buying recognized and worldwide verified products (Patrzałek *et al.*, 2010). One may speak about two meaning dimensions of brand in the context of its features, benefits, values and personality. First, a brand is associated with product features. The highest quality, reliability, durability, a well-made product, considerable value and prestige are only some of the features that may be used for brand positioning. However, customers do not buy features but benefits, therefore the features should be presented as functional or emotional benefits for the buyers. The brand may express the values to which the buyers adhere and which show their personalities. It is attractive primarily for buyers for whom the desired or real image complies with the brand image (Kotler *et al.*, 2002).

Global brands are associated by buyers with their availability on various markets and a global image. The image is composed of their universality, quality, prestige but also social responsibility (Badowska, 2014) (Table 12.4). The difficulty that enterprises face while creating global brands is the necessity to convey the values represented by the brand to many national markets, which in most cases are not homogenous concerning the preferences of local customers (Merino and Gonzalez, 2008).

Due to its brand, a global product strongly influences the consumer. In a global product, the customer's interests are usually shifted from the basic usable functions to visual features of the product perception. A global product more often exposes the emotional value than material value of the product.

Table 12.4. Characteristics of global brands (Badowska, 2014).

User	Cosmopolitan, globalist, man of the world
User personality traits and attitudes	Extravert, joyful dynamic, modern, postmodernist, proud of holding universal values, recognizing the similarities, not differences between people in the world, interested in the events on a global scale, identifies himself with the people from all over the world
Brand concept	Primacy of homogenous needs and preferences of buyers on many markets
Name and symbol	Refers to a specific product or firm, the name sounding identically in national languages
Reputation	High loyalty, trust
Recognition	Brand recognized in the same way on various target markets
Functionality and emotions	Innovation and standardization emphasized in the area of functionality. Brand referring to universal values and standards held commonly by a majority of people in the world, such as: joy, love, happiness and success
Characteristics	Universalism, standardization, high quality, brand features adjusted to the needs and preferences of wide target groups
Benefits	Including in the global trends, prestige, global image
Values	Universal for the mankind, cosmopolitan, liberal, accepted by general community, common for the human population
Culture	International, open, democratic

Brand globalization processes take place as a result of processes of lifestyle convergence and the occurrence of global consumer segments. Young consumers with positive attitudes to modern technologies, new media and values associated with globality and universal communication are more open to the conception of global brands (Żak, 2009).

However, not all buyers are equally susceptible to the influence of global brands. The attitude of buyers towards global brands allows to identify four main types, i.e. global citizens, global dreamers, antiglobalists and global agnostics. The first segment comprises consumers sensitive to brand 'globality', expecting pro-social behaviours from the brand. The second segment groups people who identify global brands with widely disseminated myths about them. They expect that while buying global brands they will become citizens of the world. In contrast, antiglobalists do not value global brands. They avoid purchasing them. In global agnostics' opinion, the sole fact that a given brand is global is not interesting, it does not guarantee the high quality of the goods (Bachnik, 2007).

Marketing of a majority of global brands differs in order to fulfil the preferences and needs of consumers. Even though a company promotes its brand all over the world, it is difficult to unify local associations in various countries. The brand is a promise of a set of determined benefits for the consumers, therefore the possibilities of adaptation are very wide. Several methods of brand creation and acquisition by an enterprise on an international market may be indicated (Altkorn, 2003).

The enterprise may introduce a brand that has been using so far on the domestic market to a foreign market. This involves the transfer of its strategy and values to the new market. Such a brand must possess characteristics that should be universal and interesting for consumers on new target markets. Since all marketing operations on foreign markets are strongly conditioned by culture, the fundamental factor of the success of this strategy is the absence of negative connotations with the brand name and its image on a given market. The enterprises that have resources enough may decide to simultaneously introduce their brand to many markets (the 'shower' model). On the other hand, enterprises with limited resources may decide for a gradual entering of foreign markets (the 'waterfall' model) (Limański and Drabik, 2010). The advantage of this strategy is no necessary brand adaptation to the local conditions, which provides an opportunity to create a brand with global reach (Duliniec, 2007).

Another way to develop a global brand is creating and developing so called global platform, supplemented with local, adapted elements, following the rule: 'think globally, act locally'. The strategy is used e.g. in the whole fast food sector (Altkorn, 2003).

The next approach to international brand building is the identification of global needs and requirements and developing a new brand to fulfil them. This strategy leads to creating completely new, previously unknown world markets. The autonomous building of a new brand is justified if the enterprise operates on an expansible market, has a product with unique features or a totally innovative one. The process of building a new brand on the international market requires time, it is complicated and risky. It involves creating new identifiers, such as the name, logo, colours, slogans and then building the brand position in the awareness of consumers (Limański and Drabik, 2010). The autonomous building of new brands is a decision that requires considerable material resources and competences in the area of marketing, because the strategy is quite risky. Many new brands created in this way are rejected and are not accepted either at the stage of market testing or during commercialization.

Still another way of brand creation on international markets is a takeover of a local brand through a purchase or merger and its subsequent internationalization. This strategy is commonly used on traditional slowly developing markets, where the competition is strong. Therefore, creating a new brand by an enterprise may be expensive and time consuming, whereas introducing a brand so far used on other markets is burdened with the risk of rejection. In view of the above, enterprises decide for the acquisition (purchase) of the brand, which allows for a fast entering of the foreign market. Local, regional and even brands with global outreach are acquired in this way (Limański and Drabik, 2010). This strategy is applied by many companies in the food sector (among others Coca-Cola, Nestle, Danone, PepsiCo, Kraft Foods), who buy strong local brands also in order to widen the product range for a given market. The tendency is also linked to an increasing orientation of global companies towards specific features of local markets and an attempt to reach those buyer segments who are not interested in the brands previously available from the portfolio of a global firm. Simultaneously, the acquisition of a local brand does not prevent introducing also the brand previously used by the enterprise to the new market (Duliniec, 2007).

Enterprises may also use a so-called multi-local strategy, which involves using various names of brands in different countries. A change of name may be enforced by the legal regulations, such as an act on language protection, but also by political or cultural requirements. It may be also due to language reasons. Moreover, certain symbols and values held by nations may differ, even in case of neighbouring countries, whose dwellers have a different sense of humour, different values, but also different attitudes towards the product and its use (Mikołajczyk, 2010). Algida uses the strategy of product individualization, adjusting to economic, cultural and technological conditions on individual markets. In Poland, the Algida brand operates under the same name as Langnese in Germany, as Ola in Spain, Belgium and Holland – and Walls's in Great Britain (Adamczyk and Witek, 2008).

Appraisals and rankings of global brands have been conducted for many years. According to the Interbrand consulting company, which annually conducts research on brand value, a recognized brand may constitute up to 70% of the firm market value. Strategic decisions concerning brands may be of key importance for its success on foreign markets. In 2015, Interbrand presented its subsequent 'Best Global Brands' report, i.e. 100 of the most valuable brands in the world. The ranking of the top rated brands in the food sector were as follows: Coca-Cola (3), McDonald's (9), Pepsi (24), Budweiser (31), Kellog's (34), Nescafe (36), Danone (51), Nestle (52), Starbucks (67), KFC (75), Sprite (81), Jack Daniels (84), Heineken (89), Johnnie Walker (92) and Smirnoff (94). For many years the Coca-Cola brand was regarded as the most precious brand in the world. Currently the two top positions in the ranking belong to brands from the new technologies sector (Apple and Google).

12.5 Globalisation of consumption

Consumption is one of the main conditions for the development of society and it plays a dominant role in comparison with other spheres of social life. Globalization processes are not without influence on changes of the consumer values system. Globalization is treated both as the cause and effect of changes occurring in consumer behaviours. Globalization of consumption involves the spreading of identical or similar consumption patterns over the national scale (homogenization of consumption) and creating so called global consumer culture. A more detailed analysis allows three main areas of globalization of consumption to be identified (Olejniczuk-Merta, 1999):

▶ Market environment and the conditions determining fulfilment of consumer needs (increase in the importance of super- and hypermarkets, trade and services infrastructure, popularization of self-service and legal regulations for the consumption sphere and market, in the first place including the laws protecting consumer interests).
▶ Ways of fulfilling needs, purchasing habits and buyers' and consumer behaviour in the individual consumer segments (including the product range and brands of chosen commodities and consumer services).
▶ Hierarchy and structure of the buyers' needs, their systems of values, attitudes and lifestyles.

The demand side of the factors favouring globalization is represented by the consumers, their strivings and expectations (Table 12.5). The success of global strategy is determined by meeting the expectations of various consumer groups through creating a global product. Among the demand stimuli favouring globalization of consumption, the consumer striving towards a modern, more attractive lifestyle is the most visible. Consumers seek convenient products of a high quality, which in their mind are represented by a global product – accepted in various parts of the world. Buying global products is also a way to boost consumers' self-esteem, particularly when it concerns consumers from less developed countries. Global products express an attractive, modern and desired lifestyle; they carry and are a symbol of certain system of values and philosophy of life of contemporary consumers. Purchasing global products also facilitates communication among consumers from different countries due to the similarity of consumption patterns (Mazurek-Łopacińska, 2003).

Considering the supply side, globalization of consumption is stimulated by the development of competition, which requires application of a strategy of entering new markets and seeking financially advantageous operating conditions on these markets by achieving economies of scale. On the other hand, global strategies would have never gained such spectacular success without the development of information technologies, which allow a specific link with the market. Globalization of consumption is also accelerated by the development of the world wide web, breaking the barriers to information access. By acquiring more and more information about a product, a consumer must have equally easy access to these goods.

The main factor inhibiting globalization of consumption is cultural conditions. Culture fulfils an important function in the sustainable development of societies; people draw ideas, values and symbols crucial for their development from it. However, the free flow of the workforce and of capital characteristic for globalization leads to the creation of nations beyond their native lands, because immigrants form cultural enclaves in a foreign land.

Table 12.5. Factors affecting globalization of consumption (Mazurek-Łopacińska, 2003).

Factors driving globalization of consumption		Factors inhibiting globalization of consumption
Demand	Supply	
Consumer aspiration for a modern, more attractive lifestyle	Growing competition enforcing application of strategy of expansion into the new markets	National attitudes causing a preference of domestic products
Seeking convenient global products of high quality	Enterprises striving to reach economies of scale	Cultural conditions
Buying global products as a way to boost consumer self-esteem in the less developed countries	Free flow of products, workforce and capital	Income stratification
Increased spatial mobility of consumers	Development of information technologies ensuring efficient communication with the market	Unemployment

Globalization leads to mixing cultural patterns. Cultural diversity is of crucial importance in the globalization process, it affects its course and forms. Respect for cultural difference favours the acceptance of world, global products by a consumer (Bogunia-Borowska and Śleboda, 2003).

The effect of the globalization process on consumer behaviours creates an image of global consumers, perceiving himself as a person who prefers modern consumption patterns and combining a traditional system of values with new patterns of behaviour (Patrzałek *et al.*, 2010). Consumers buy global products for various reasons. The most important include the following (Mazurek-Łopacińska, 2001):

- Global products express a modern and attractive lifestyle, desired by many buyers, moreover they are the carrier and symbol of the system of values and philosophy of life of contemporary consumers.
- Global products are usually characterized by functionality, high quality, they are convenient and useful.
- Using a global offer facilitates communication among consumers from different countries.

The assessment of globalization of consumption processes is not unanimous. It generates both positive and negative results, the scale and range of which depend on the level of a country's socio-economic development (Table 12.6).

Table 12.6. Outcomes of globalization of consumption (Żmija *et al.*, 2010).

Positive	Negative
• spreading the ethics of global consumption	• creolization – artificial import of foreign influence and products to local tradition
• wider accessibility of high technology products	• risk of changing local consumer habits
• better terms of meeting consumer needs and expectations,	• threats to consumers in poorly developed countries due to insufficient information and education
• better life standard	
• strengthening consumer preferences	• dangers resulting from acquisition of global products at the cost of fulfilling basic needs
• unification of legal regulations on consumer safety	• lack of correlation between market globalization and globalization of democratic and citizen institutions
• development of infrastructure	
• acceleration of interpersonal communication (transfer of information technologies to everyday life)	

The benefits result from the unification of products and their way of use on the world scale, disregarding local cultures, climate, etc. These phenomena often make life easier and increase human spatial mobility, particularly tourism and migrations.

One of the most positive outcomes of globalization is the wide dissemination of ethics of global consumption that establishes certain consumer behaviours. It is expressed not only in the acquisition of goods recognized worldwide, but also in using services allowing to identify oneself with model-creating consumer groups. The features of global consumption ethics comprise purchasing of goods recognizable worldwide, consumer awareness of sustainability, sensitivity to harm done to the environment in the course of product utilization, increased interest in pro-environmental campaigns, increased consumer interest in problems of health and care for the proper functioning of their organisms, acceptance of universal human values (Mazurek-Łopacińska, 2003).

Globalization of consumption also has negative outcomes. Generally, it inhibits the development of local cultures and destroys local products and ways of consumption, which may lead to cultural impoverishment of local communities over a longer period of time.

References

Adamczyk, J. and Witek, L., 2008. Marketing międzynarodowy. Oficyna Wydawnicza Politechniki Rzeszowskiej, Rzeszów, Poland, 188 pp.

Altkorn, J., 2003. Podstawy marketingu. Instytut Marketingu, Kraków, Poland, 437 pp.

Bachnik, K., 2007. Konsument wobec marek globalnych. In: Kusińska, A. (ed.) Konsumpcja a rozwój gospodarczy. Instytut Rynku Wewnętrznego i Konsumpcji, Warszawa, Poland, pp. 140-147.

Badowska, S., 2014. Istota i tożsamość marek lokalnych i globalnych. Marketing i Rynek 8: 320-325.

Bogunia-Borowska, M. and Śleboda, M., 2003. Globalizacja i konsumpcja – Dwa dylematy współczesności. Universitas, Kraków, Poland, 316 pp.

Czerny, M., 2005. Globalizacja a rozwój. Wybrane zagadnienie geografii społeczno-gospodarczej świata. Wydawnictwo Naukowe PWN, Warszawa, Poland, 240 pp.

Drabik, I. and Limański, A., 2010. Problemy kształtowania produktu globalnego w strategii marketingowej przedsiębiorstwa. Zeszyt Naukowe Akademii Ekonomicznej w Poznaniu 153: 35-41.

Drachal, K., 2014. What do we know from EPRG model? EcoForum 3(2): 85-92.

Duliniec, E., 2007. Marketing międzynarodowy. Uwarunkowania, instrumenty, tendencje. Szkoła Główna Handlowa, Warszawa, Poland, 352 pp.

Dybowski, G., 2005. Globalne regulacje w światowym handlu żywnością. Ocena wpływu na rozwój rolnictwa na świecie. In: Dybowski, G. (ed.) Wpływ procesu globalizacji na rozwój rolnictwa na świecie. Instytut Ekonomiki Rolnictwa i Gospodarki Żywnościowej, Warszawa, Poland, pp. 40-60.

Fonfara, K., Deszczyński, B., Dymitrowski, A., Małys, Ł., Mielcarek, P., Ratajczak-Mrozek, M., Nowacki, F., Soniewicki, M. and Wieczerzycki, M., 2014. Marketing międzynarodowy. Współczesne trendy i praktyka. PWN, Warszawa, Poland, 241 pp.

Friedland, W.H., 2003. Agrifood globalization and commodity systems: the globalization of agriculture and food at the agriculture and human values society. Available at: http://tinyurl.com/y9c6oy47.

Goryńska-Goldman, E., 2009. Ewolucja zwyczajów żywieniowych i ich znaczenie obecne. Marketing i Rynek 12: 18-24.

Grunert, K.G. and Valli, C., 2001. Designer-made meat and dairy products: consumer-led product development. Livestock Production Science 12(1-2): 83-98.

Grzegorczyk, W., 2005. Marketing na rynku międzynarodowym. Oficyna Ekonomiczna, Kraków, Poland, 232 pp.

Hollensen, S., 2007. Global marketing. A decision-oriented approach. Pearson Education Ltd., London, UK, 714 pp.

Horská, E., 2014. Specifics of international marketing and management in Visegrad Countries – Qualitative analysis of selected case studies. In: Durendez, A. and Wach, K. (eds.) Patterns of business internationalization in Visegrad Countries – In search for regional specifics. Universidad Politecnica de Cartagena, Cartagena, Spain, pp. 161-178.

Horská, E., Prokeinova, R., Galova, J., Kadekova, Z., Krasnodębski, A., Maitah, M., Matysik-Pejas, R., Paluchova, J., Nagyova, L., Omarkulova, M., Pribyl, M., Smutka, L., Szabo, Z. and Wach, K., 2014. International marketing. Within and beyond Visegrad borders. Episteme, Krakow, Poland, 311 pp.

Horská, E., Ubreziova, I. and Kekäle, T., 2007. Product adaptation in processes of internationalization: case of the Slovak food-processing companies. Baltic Journal of Management 2(3): 319-333.

Kall, J., Kłeczek, R. and Sagan, A., 2006. Zarządzanie marką. Oficyna Ekonomiczna, Kraków, Poland, 312 pp.

Kotler, P., Armstrong, G., Saunders, J. and Wong, V., 2002. Marketing. Podręcznik europejski. Polskie Wydawnictwo Ekonomiczne, Warszawa, Poland, 625 pp.

Lenart, A., 2008. Projektowanie nowych produktów spożywczych. Przemysł Spożywczy 4: 2-7.

Liberska, B., 2002. Globalizacja – Mechanizmy i wyzwania. Polskie Wydawnictwo Ekonomiczne, Warszawa, Poland, 342 pp.

Limański, A. and Drabik, I., 2010. Marketing międzynarodowy. Difin, Warszawa, Poland, 415 pp.

Mazurek-Łopacińska, K., 2001. Globalizacja w aspekcie wpływu na zachowania konsumenckie. Marketing i Rynek 3: 10-16.

Mazurek-Łopacińska, K., 2003. Zachowania nabywców i ich konsekwencje marketingowe. Polskie Wydawnictwo Ekonomiczne, Warszawa, Poland, 370 pp.

Mazurek-Łopecińska, K., 2015. Uwarunkowania marketingowej strategii produktu w warunkach globalizacji. Available at: http://tinyurl.com/y9y43sc6.

Merino, M. and Gonzalez, S., 2008. Global or local? Consumers' perception of global brands in Latin America. Latin American Advances in Consumer Research 2: 16-21.

Mikołajczyk, A., 2010. Strategia adaptacji marki globalnej do rynków lokalnych. Studia Gdańskie 7: 9-23.

Naisbitt, J., 1997. Mega trendy: dziesięć nowych kierunków zmieniających nasze życie. Wydawnictwo Zyska i S-ka, Poznań, Poland, 304 pp.

Oczkowska, R., 2006. Product strategies on foreign markets and determinants of their selection. Zeszyty Naukowe Akademii Ekonomicznej w Krakowie 725: 37-50.

Olejniczuk-Merta, A., 1999. Zachowanie młodych konsumentów w Polsce w warunkach globalizacji rynków. In: Karcz, K. (ed.) Konsument i przedsiębiorstwo w przestrzeni europejskiej. V Międzynarodowa Konferencja Sieci Krajów Grupy Wyszehradzkiej, Katowice, Poland, pp. 70-79.

Olejniczuk-Merta, A., 2014. Refleksja nad zmianami w konsumpcji i zachowaniu konsumentów w erze globalizacji. Marketing i Rynek 8: 486-494.

Patrzałek, W., Banaszczak-Soroka, U., Bażański, E., Cebula, M., Kucharski, P., Majkut, R., Miklaszewski, L., Orzeszyna, J., Perchla-Włosik, A., Srokowski, Ł., Wardzała-Kordyś, J., Wilk, A., Wójcik, D. and Zamiar, A., 2010. Zachowania podmiotów w warunkach globalizacji rynków. Wydawnictwo Naukowe Scholar, Warszawa, Poland, 303 pp.

Sudoł, S., Szymczak, J., Haffer, M., Andruszkiewicz, K., Haffer, R., Sudolska, A. and Szumowski, T., 2000. Marketingowe testowanie produktów. Polskie Wydawnictwo Ekonomiczne, Warszawa, Poland, 356 pp.

Sztucki, T., 1998. Marketing w pytaniach i odpowiedziach. Placet, Warszawa, Poland, 271 pp.

Szymański, W., 2004. Interesy i sprzeczności globalizacji. Wprowadzenie do ekonomii ery globalizacji. Difin, Warszawa, Poland, 374 pp.

Taranko, T., 2015. Strategie nowego produktu i związane z nimi ryzyko. Zarządzanie produktem w warunkach globalizacji gospodarki. Akademia Ekonomiczna, Poznaniu, Poznań, Poland.

Usunier, J.-C. and Lee, J., 2005. Marketing across cultures. Prentice Hall, Upper Saddle River, USA, 496 pp.

Vrontis, D. and Thrassou, A., 2007. Adaptation vs standarization in international marketing – The country-of-origin effect. Innovative Marketing 3(4): 7-20.

Wach, K., 2003. Produkt globalny jako jeden z efektów procesu globalizacji. Prace Naukowe Akademii Ekonomicznej we Wrocławiu 976: 411-422.

Wierzbicka, B., 2012. Standaryzacja i adaptacja strategii – Dylemat firm działających na rynkach zagranicznych. Zarządzanie i Finanse 2: 177-190.

Wiktor, J.W., 2006. Company strategies in international marketing. Towards typology. Zaszyty Naukowe Akademii Ekonomicznej w Krakowie 729: 17-36.

Wiktor, J.W., Oczkowska, R. and Żbikowska, A., 2008. Marketing międzynarodowy. Zarys Problematyki. Polskie Wydawnictwo Ekonomiczne, Warszawa, Poland, 348 pp.

Yip, G.S., 2004. Strategia globalna: światowa przewaga konkurencyjna. Polskie Wydawnictwo Ekonomiczne, Warszawa, Poland, 334 pp.

Żak, A., 2009. The product in global activities of enterprises. Zeszyty Naukowe Wyższej Szkoły Ekonomii i Informatyki w Krakowie 5: 119-133.

Zięba, K., 2010. Cultural influence on consumer behaviour. Zeszyty Naukowe Uniwersytetu Szczecińskiego. Problemy Zarządzania, Finansów i Marketingu 16: 401-411.

Żmija, J., Matysik-Pejas, R. and Szafrańska, M., 2010. Globalization of the consumption – Determinants and effects. DSM Business Review 2(1): 43-64.

13. Success factors in new product development in the food sector

R. Matysik-Pejas

Agricultural University in Kraków, Al. Mickiewicza 21, 31-120 Kraków, Poland; rrmatysi@cyf-kr.edu.pl

Abstract

The food product market is a special market. Its importance is derived from the fact that the products it offers constitute one of the most important group of consumption goods for humans. The competition of food markets causes food manufacturers to undertake activities aimed at increasing their competitiveness and distinguishing their market offering, among others by marketing innovative food products. The market success of new products is determined by a number of the company's internal conditions, but also but by external factors, which must be taken into consideration when developing and marketing new products. Among these issues are consumer needs and their evolution visible in new consumer trends. Food companies should focus their attention on existing and forming trends. They may also cooperate with consumers in the process of new product development. Market success of new food products is also determined by the kind and range of marketing activities used by the company. Companies must remember that whether new products will stay in the market is determined by the extent to which they will be accepted by the final consumers.

Keywords: food, new product, success factors

13.1 Introduction

Innovation is regarded as one of the basic functions of contemporary enterprises, determining their development. Innovations are introduced in many interrelated areas of the company's activities. Product innovations are the result of the enterprise's innovation activity, visible on the market. The efficiency of developing and launching new products depends on competences, experiences, resources, method of management, but also on the marketing strategies adopted by the enterprise (Fortuin *et al.*, 2007; Grunert and Sorensen, 1996; Suwannaporn and Speece, 2003). The attributes of the product itself, as high quality, novelty of product, price, brand and other are extremely important (Cooper and Kleinschmidt, 2000). One of the key issues concerning new food product development are market and consumer oriented activities (Costa and Jongen, 2006; Earle, 1997; Stewart-Knox and Mitchell, 2003). Both consumers and their market environment are undergoing transformations. The changes become visible in the changes in previous needs structures and the formation of new needs, in the systems of values, attitudes and lifestyles, and in the conditions for the fulfilment of needs, ending with purchasing and consumption behaviours. Enterprises must consider many of these factors in their decisions on new product development and must implement them better than their competitors. Innovations accepted by the market may contribute to gaining a competitive advantage by an enterprise and even help to gain a periodic leadership position on a given product market.

13.2 Consumer trends and their importance for the success of new food products

Consumer needs change constantly, therefore products which were in great demand recently may lose their popularity among consumers in the future (Linnemann *et al.*, 2006). Enterprises that introduce product innovations to food markets should analyse the environment and changes that occur in consumer trends. Time plays a crucial role in making use of appearing trends in a marketing strategy. The accurate anticipation of newly forming trends and timely (ahead of the competition) implementation to create a product offer may mean measureable benefits and gaining a market advantage by the enterprise over a longer period of time (Mróz, 2013).

The innovative trends in food manufacturing, which may currently be distinguished in the European food industry, are grouped into main axes, meeting consumer demands. These trends are connected with health, convenience, sustainability, authenticity, and pleasure (Figure 13.1).

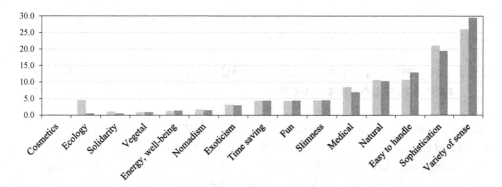

Figure 13.1. Food innovation trends in Europe in year 2011 and 2014 (%) (Data and Trends European Food and Drink Industry, 2012; Data and Trends European Food and Drink Industry, 2014-2015).

The enterprises of the agri-food sector should focus their attention on the correct analysis of the existing and still forming trends. The knowledge about consumer behaviours is supplied by all kinds of statistical data concerning population, reports and results of empirical research focused on lifestyles, etc. The knowledge about consumer behaviours has a practical dimension and may be used in the process of developing marketing strategies in enterprises. The familiarity with trends allows the needs and motivations of customers to be foreseen and may prove to be a driver of success in the process of new product marketing and the creation of new markets. This requires anticipation of consumer needs well in advance and enforces an on-going monitoring of the environment. Other circumstances also result in knowledge about the tendencies in consumer behaviour being important for the future success of product innovations on the market (Mróz, 2013):

- changes take place ever faster and new product generations displace the previous ones;
- innovative distribution channels and manners of promotion are developing;
- changes in the systems of values and consumer lifestyles occur faster than demographic changes of population;
- consumers have increasingly greater influence on developing the product offer and opinions about them;
- globalization and greater mobility of consumers removes the barriers in the diffusion of new trends worldwide;
- consumers become increasingly more demanding towards enterprises and their offers.

Monitoring and identifying new consumer trends is usually the task of specialized research agencies. However, enterprises using such studies should remember that the professional value and analytical accuracy of such reports may vary. They do not always reflect the significant phenomena and social processes and often focus on giving media publicity to alleged new trends (Mróz, 2015).

13.3 Cooperation of enterprises with consumers as a factor of new food product success

The main aim of marketing innovative products is fulfilling current or arising consumer needs and expectations. Consumers decide whether the products will find acceptance on the market. To achieve success, the innovative products must be developed and tailor-made for consumers. Consumer orientation is one of the major drivers of success in the food manufacturing enterprise (Fortuin and Omta, 2009; Fortuin *et al.*, 2007). Consumers play a key role at each stage of the new product development process (Czajkowska *et al.*, 2013). Their participation may be passive or active. The passive function means that enterprises, including a new product in their offer, use their knowledge about consumer behaviour for this purpose. On the other hand, the active function is a form of direct consumer involvement in the new product development process (Sojkin *et al.*, 2012).

Depending on the degree of consumer involvement in the company's operations three types of active consumers may be distinguished. The first group consists of active consumers, whose involvement is nonetheless limited only to the assessment and evaluation of products on the internet forums or the websites of the stores. The second group consists of the consumers participating in activities which purpose is to make the offer more attractive as well as the consumers taking part in the campaigns prepared by the companies. The third group constitutes the innovative prosumers, who exhibit the aforementioned actions and try to influence the companies' offer by themselves (Szul, 2013).

Initiatives of enterprises concerning product development with consumer participation can build on different concepts. An example may be open innovation and open source concepts. Open innovation involves the exchange of ideas between the enterprise and its environment (among others consumers). Despite the fact that there are few examples of the widespread use of open innovation by enterprises in the food sector, its potential is considerable. Perhaps this idea will be appreciated in the future by the food sector (Fortuin *et al.*, 2009). The other idea mentioned above, i.e. open source, is based on free sharing and use of ideas. An example may be the case of a Canadian company, which made its recipe for a cola type carbonated beverage available to consumers. Consumers who applied this recipe could produce the beverage and freely modify the recipe. The precondition for making it available was the requirement to present a modified version of the recipe (Czapski, 2012).

In the product commercialization process consumers can actively participate in planning the marketing activities connected with the marketing of innovations. They can be involved in planning the activities concerning the final refinement of the packaging. Their opinions can also be taken into consideration in creating messages in market communication, so that it reaches their target market segments (Sojkin *et al.*, 2012). Moreover, consumers may be helpful in the choice of the manner of distribution for the new product manufactured by the enterprise. The role of the consumer is also valuable when the level of the product's market price is to be established. Taking their price sensitivity into consideration may significantly contribute to the market success of the newly introduced product.

13.4 Level of innovation in food products

When determining the drivers of a new product's success, one should mention the considerable role of the level of innovation. The level of innovation depends on the fact whether the new product is a copy or a slight modification of previous products, or a complete novelty (Earle *et al.*, 2007). Enterprises usually are very cautious and most often market products that are variations of previously offered products. However, even slight alterations in the product may bring new benefits to its buyers. These products may differ from competitive products by a particular feature or be of a similar quality as more expensive products, fulfilling the same needs (Pasternak, 2001). However products constituting a line extension and me-too products bring only short-term profits (Fortuin and Omta, 2009). But it is also a known fact that a radical innovation is associated with a much stronger risk and may result in failure, because it is impossible to predict the market's reaction.

Considering the extent of potential innovative changes in food products, the fact that it is difficult to develop totally new food products (in comparison with other industries) should be taken into consideration. So far, revolutionary changes have been few and distant in time. Many consumers treat new food products cautiously and with mistrust, because many consumers change their habits only slowly, unless they perceive the benefits resulting from the new product. Consumers most quickly accept new food products when the risk involved in their consumption is little, when they have much better properties than the products already available on the market, are not a total novelty, fulfil consumer needs and desires, can be tested earlier, are intensively advertised and promoted and are convenient to use (Garbarski, 1998).

13.5 New product packaging as a driver of market success

As it has already been mentioned, the market success of a new food product is to a considerable extent affected by its packaging. It is an integral part of the product and plays the role of a silent seller. The packaging is a means of communication with the market, increases the interest of customers and may make them realize their needs. A well designed packaging should be an inseparable element of a new product.. In self-service shops, packaging not only protects the product, but also fulfils the information and promotional function. Therefore, the aesthetic look (properly chosen colours and shapes), information put on it (composition, way of preparation), size and practicality (ease of use, convenient consumption, storage) are extremely important for the new product. Frequently, while shopping, the buyers are attracted by the look of the packaging and on this basis imagine the product and its characteristics. When a new product appears on the market, it is the packaging that usually determines the first purchase by a consumer.. Therefore, packaging should draw consumer's attention, highlight the product value and differentiate it from other similar products (Pasternak, 2001). The promotional function of the new product packaging is a tool to stimulate demand, encourage the consumer to unknowingly pay attention and to make a decision to buy the product. While designing the new project packaging, the enterprise should adjust its promotional features to the needs and psychological, cultural or age conditionings of the buyer groups for whom the product is intended (Cichoń, 1996).

13.6 Process of consumer acceptance of new product

Consumer response to a new product is a complicated psychological process that starts with finding information about the innovation and ends with its adoption. The notion of new product adoption denotes making a decision by a consumer to buy and use it (Kramer, 2004). The process differs depending on the consumer, because new food products spark curiosity and interest in some consumers, while they evoke fear and stress in others. In addition, it should be noted that the attributes of innovations attract potential consumers to varying degrees. These are (Rogers, 1983):

- *Comparative benefits*: the degree to which a proposed innovation offers additional benefits and values in comparison with the existing products; the higher the relative advantage in consumer perception, the faster the adoption of the new product (among others improvement of sensory and healthy features of new food products, saving of time and effort while preparing new foodstuffs).
- *Compatibility*: the better the compatibility between the innovation and the existing standards, consumer experiences and needs, the faster the innovation diffusion process (among others the compatibility of new food products with consumption standards and consumer lifestyles).

▶ *Complexity*: the more complex the innovation, the harder to understand or use, the more difficult it will be to encourage consumers to adopt it, so the diffusion process will be slower (e.g. too complicated method of new kind of food preparation).

▶ *Possibility to test the innovations*: possibility to test the innovations by consumers 'as a trial' reduces their feeling of indecision or risk and therefore facilitates diffusion of innovations.

▶ *Perceptibility of innovations*: if the results of the adoption of innovation are clearly visible, the innovation will be adopted more readily and faster (among other perceiving the advantages of consuming new kinds of food, e.g. functional or organic food).

Despite the fact that innovative products may bring consumers numerous benefits, their behaviours evidence some reluctance to accept them. The resistance results in the market adoption process of the new product slowing down. Consumer opposition against new products may assume various forms from postponing the decision about the purchase to the most extreme case, i.e. their rejection (Cornescu and Adam, 2013). New food is associated with the term 'food neophobia', i.e. a general aversion to consume new food products (Babicz-Zielińska *et al.*, 2009). As mentioned previously, the emotions caused by new foods are linked to the level of innovation. A small alteration of the product may spark interest and a desire to learn, whereas a total novelty may lead to negative emotional states, such as apprehension or nervousness. Negative emotions result from unfamiliarity with the product. Resentment of new products is connected with human physiology and the emotional sphere. For instance, timid persons, who do not like the risk are more reluctant to consume new food (Babicz-Zielińska *et al.*, 2009). Consumer inclination towards getting interested in new food products and their purchase is also correlated to their customs and gustatory habits, which are the main factors affecting their purchasing decisions. It constitutes a type of sensory, but also hedonistic barrier, which the food manufacturers face (Gutkowska, 2011).

Consumers who are exposed to new products are able to perceive both the risk and benefits that the new product gives. Consumer attitudes towards new products result from previous experiences, but also from the opinions of other consumers. Some consumers are risk-takers, unafraid of testing new products, others are more cautious and will not dare to taste a new product without exhaustive information about it. For efficient dissemination of novelties on the food market, not only the needs and expectations of potential buyers should be known, but also ways to encourage them to test and buy the new product (Earle *et al.*, 2007). An important element of consumer education concerning new products is an intensive process of their promotion, aiming at breaking the dissonance between the individual components of consumer attitudes towards food. On the one hand it aims to inform consumers about new products, on the other to create positive confidence and in result forming appropriate purchasing behaviours. Avoiding new, unknown food products by consumers may makes the manufacturers undertake promotional activities, e.g. product tasting at the point of sale. It can help to break the sensory barrier (Gutkowska, 2011). In this way consumers have the opportunity to taste these products before buying them, which diminishes their doubts and uncertainty.

Acceptance, i.e. adoption of an innovative product by the buyer is a process composed of several stages: realization stage, interest stage, assessment stage, testing and verification stage and adoption stage (Figure 13.2). The realization stage is when a potential buyer learns that a new innovative product has appeared on the market, but otherwise has no information about it (Garbarski, 1998). At this stage the first contact of the consumer with the new product takes place, but the knowledge and information that the buyer possesses about this product are still scarce. Contact with the new product and deficiency of information stimulate a passage to the next stage – interest stage.

During the interest stage more information about the new product reaches a potential buyer, causing growing interest (Kaczmarek *et al.,* 2005). While launching an innovative product into the market, its manufacturer should supply the information about the new product to consumers and adjust its kind to the phases of product acceptance through which a consumer passes. For instance, at the stage of creating awareness and interest, the best results are achieved using mass media advertising. In the campaign bringing an innovative product onto the market it is necessary to make potential consumers aware of the product existence in the fastest and most efficient way.

The third stage, the assessment stage (i.e. determining value) is the moment when a potential buyer considers buying the new product (Kramer, 2004). S/he thinks about the benefits

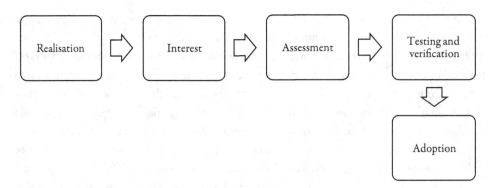

Figure 13.2. Stages of innovation acceptance (Rogers, 1983).

that the product may bring him/her, considers whether the purchased product would fulfil the needs s/he feels and starts thinking about buying the product.

Subsequently, the next stage of innovation acceptance by consumers starts – the testing and verification stage. At this stage the new product is bought for the first time. Usually the consumer buys small quantities of the product or uses a sample. In this way, the new product buyer is able to determine its values. The testing stage is an important period for the manufacturer because it provides a very good opportunity for efficient product promotion (Garbarski, 1998).

The last stage is the adoption stage. A consumer possesses necessary information about the food product innovation, has already tested it, which leads him/her to make a decision buy the new product regularly (Mazurek-Łopacińska, 2003). At this stage it may also happen, that the consumer is not satisfied after going through the previous stages, which results in rejection of the new product.

The degree to which a new product is compatible with consumer lifestyle, values and beliefs plays a crucial role in the course of the product diffusion process on the market. The characteristic features, accumulated experience and preferences encourage or discourage the acceptance of innovations. Consumer types are characterized on the basis of their psychological traits (attitude, personality, intelligence, emotionality, behaviour), demographic features (gender, age, place of residence), needs and expectations, education, acquired knowledge, propensity to avoid or to undertake risk, openness to change. However, the key element is called innovativeness, which is the propensity to accept innovations faster than other participants (Grunert and Traill, 1997; Klincewicz, 2011; Matysik-Pejas, 2012; Wang *et al.*, 2008). To be innovative does not have to mean that one buys new products at once, but consumers having this characteristic also strive to get information about the novelties on the market, which later lead to purchase of these products (Gutkowska and Ozimek, 2005). Innovative consumers are a key market segment for companies. They play an important role of opinion leaders in the success of a new product (Huotilainen *et al.*, 2006).

Diffusion of innovations also depends on the scale of social imitation, i.e. the dimension of influence which other persons have upon the buyer, spreading other people's opinions about the innovation and also the speed of learning processes preceding the novelty acceptance. The innovation diffusion process is conditioned by factors mainly connected with consumers and their consumption patterns, and the product properties. The most important of them comprise (Mazurek-Łopacińska, 2003):

- the product importance in the hierarchy of the buyer's needs (the greater the consumer interest in the new product, the greater activity he will show in seeking information about it, which in effect will lead to an easier acceptance of innovation);
- the degree of new product acceptance by the buyer's environment (high level of appreciation for the innovative product causes positive reactions among consumers and may accelerate the purchase);
- the product price compared with the consumer purchasing power (too high a price may prove to be a barrier for consumers in making a purchasing decision);
- innovative product traits causing them to stand out on the market;
- kind of innovation and its perception by consumers;
- the speed of decision making by the buyer;
- buyer's risk propensity.

Summing up: the innovation diffusion process is complex and determined by many factors. The better the knowledge of these factors by the enterprise which introduces innovative products, the better chances for the market success of the product.

References

Babicz-Zielińska, E., Jeżewska-Zychowicz, M. and Laskowski, W., 2009. Konsument na rynku nowej żywności – Wybrane uwarunkowania spożycia. Wydawnictwo Szkoły Głównej Gospodarstwa Wiejskiego, Warszawa, Poland, 187 pp.

Cichoń, M., 1996. Opakowanie w towaroznawstwie, marketingu i ekologii. Zakład Narodowy Imienia Ossolińskich, Kraków, Poland, 222 pp.

Cooper, R.G. and Kleinschmidt, E.J., 2000. New product performance: what distinguishes the star products. Australian Journal of Management 25: 17-45.

Cornescu, V. and Adam, C.-R., 2013. The consumer resistance behavior towards innovation. Procedia Economics and Finance 6: 457-465.

Costa, A.I.A. and Jongen, W.M.F., 2006. New insights into consumer-led food product development. Trends in Food Science and Technology 17: 457-465.

Czajkowska, K., Kowalska, H. and Piotrowski, D., 2013. The role of consumer in the process of new food products design. Zeszyty Problemowe Postępów Nauk Rolniczych 575: 23-32.

Czapski, J., 2012. Opracowywanie nowych produktów żywnościowych o charakterze prozdrowotnym. Przemysł Spożywczy 66: 32-34.

Earle, M., Earle, R. and Anderson, A., 2007. Opracowanie produktów spożywczych. Podejście marketingowe. Wydawnictwo Naukowo-Techniczne, Warszawa, Poland, 381 pp.

Earle, M.D., 1997. Innovation in the food industry. Trends in Food Science and Technology 8: 166-175.

Fortuin, F.T.J.M. and Omta, S.W.F., 2009. Drivers and barriers to innovation in the food processing industry continued. A comparison of the Netherlands and the Shanghai region in China. In: Proceedings of the 3rd International European Forum on System Dynamics and Innovation in Food Networks. February 16-20, 2009. Innsbruck-Igls, Austria, pp. 483-498.

Fortuin, F.T.J.M., Batterink, M. and Omta, S.W.F., 2007. Key success factors of innovation in multinational agrifood prospector companies. International Food and Agribusiness Management Review 10(4): 1-24.

Garbarski, L., 1998. Zachowania nabywców. Polskie Wydawnictwo Ekonomiczne, Warszawa, Poland, 183 pp.

Grunert, K. and Traill, B., 1997. Product and process innovation in the food industry. Springer, New York, USA, 242 pp.

Grunert, K.G. and Sorensen, E., 1996. Perceived and actual key success factors: a study of the yoghurt market in Denmark, Germany and the United Kingdom. Centre for Market Surveillance, Research and Strategy for the Food Sector, Aarhus, Working Paper 40, 38 pp.

Gutkowska, K. and Ozimek, I., 2005. Wybrane aspekty zachowań konsumentów na rynku żywności-kryteria zróżnicowania. Wydawnictwo Szkoły Głównej Gospodarstwa Wiejskiego, Warszawa, Poland, 230 pp.

Gutkowska, K., 2011. Innowacyjność konsumentów wobec produktów żywnościowych jako warunek rozwoju rynku żywności. Konsumpcja i Rozwój 1: 108-119.

Huotilainen, A., Pirttilä-Bäckman, A.M. and Tuorila, H., 2006. How innovativeness relates to social representation of new foods and to the willingness to try and use such foods. Food Quality and Preference 17: 353-361.

Kaczmarek, J., Stasiak, A. and Włodarczyk, B., 2005. Produkt turystyczny. Pomysł – Organizacja – Zarządzanie. Polskie Wydawnictwo Ekonomiczne, Warszawa, Poland, 389 pp.

Klincewicz, K., 2011. Dyfuzja innowacji. Jak odnieść sukces w komercjalizacji nowych produktów i usług. Wydawnictwo Naukowe Wydziału Zarządzania Uniwersytetu Warszawskiego, Warszawa, Poland, 170 pp.

Kramer, T., 2004. Podstawy Marketingu. Polskie Wydawnictwo Ekonomiczne, Warszawa, Poland, 214 pp.

Linnemann, A.R., Benner, M., Verkerk, R., Van Boekel, M.A.J.S., 2006. Consumer-driven food product development. Trends in Food Science and Technology 17: 184-190.

Matysik-Pejas, R., 2012. Innowacje na rynku produktów żywnościowych i ich oddziaływanie na nabywców. Handel Wewnętrzny 3: 140-149.

Mazurek-Łopacińska, K., 2003. Zachowania nabywców i ich konsekwencje marketingowe. Polskie Wydawnictwo Ekonomiczne, Warszawa, Poland, 370 pp.

Mróz, B., 2013. Konsument w globalnej gospodarce. Trzy perspektywy. Oficyna Wydawnicza Szkoły Głównej Handlowej, Warszawa, Poland, 300 pp.

Mróz, B., 2015. Trendy konsumenckie: implikacje marketingowe i wyzwania badawcze. In: Dąbrowska, A. and Wódkowski, A. (eds.) Badania marketingowe. Praktyka nauce – nauka praktyce. Instytut Badań Rynku, Konsumpcji i Koniunktur, Warszawa, Poland, pp. 31-50.

Pasternak, K., 2001. Rozwój produktu w przemyśle spożywczym. Uwarunkowania, działania i strategie w przedsiębiorstwie. Wydawnictwo Instytutu Przemysłu Mięsnego i Tłuszczowego, Warszawa, Poland, 129 pp.

Rogers, E.M., 1983. Diffusion of innovations. Free Press, New York / London, USA / UK, 453 pp.

Sojkin, B., Ankiel-Homa, M., Małecka, M., Michalak, S., Olejniczak, T., Pachołek, B. and Sielicka, M. 2012. Komercjalizacja produktów żywnościowych. Polskie Wydawnictwo Ekonomiczne, Warszawa, Poland, 230 pp.

Stewart-Knox, B. and Mitchell, P., 2003. What separates the winners from the losers in new food product development? Trends in Food Science and Technology 14: 58-64.

Suwannaporn, P. and Speece, M., 2003. Marketing research and new product development success in Thai food processing. Agribusiness 19: 169-188.

Szul, E., 2013. Prosumption as the activity of modern consumers – Conditions and symptoms. Nierówności społeczne, a wzrost gospodarczy 31: 347-358.

Wang, G., Dou, W. and Zhou, N., 2008. Consumption attitudes and adoption of new consumer products: a contingency approach. European Journal of Marketing 42: 238-254.

14. Towards more open innovation in the food sector

K.G. Grunert[1]*, *D. Brohm[2] and N. Domurath[2]*

[1]Aarhus University, Fuglesangsalle 4, 8210 Aarhus V, Denmark; [2]INTEGAR – Institut für Technologien im Gartenbau GmbH, Schlüterstr. 29, 01277 Dresden, Germany; klg@mgmt.au.dk

The consumer trends described in the first part of this book – health, authenticity, sustainability and convenience/bundling – have one thing in common: they will result in more complex food products. Complexity will increase in three different areas.

First, and in spite of the fact that the authenticity trend in fact may encompass a call for simplicity, the physical food products need to live up to multiple requirements. Taste and other sensory requirements need to be made compatible with health requirements, for example by finding ways to reduce the content of fat, sugar and salt without compromising the taste. Likewise, taste and other sensory requirements need to be made compatible with demands for less use of chemicals and milder forms of processing, calling for new forms of preservation and assurance of food safety.

Second, these new food products can be brought about only by cooperation in the value chain. Traditionally, the adaptation of food to differentiated and newly developing consumer needs and wants has been dealt with in the later processing stages of the food chain, with primary production and the first processing stage concentrating on efficiency and output of homogeneous raw materials (Grunert *et al.*, 2005). However, the demands for healthy, sustainable and authentic food require a combined effort in the whole food chain. Sustainable food production involves by definition the whole food chain. Authentic food production means a value chain that is transparent, not overly complicated and that involves at all levels the use of technologies that consumers can relate to. And healthier food products start with the production of healthy raw materials at the farm level.

Third, all trends discussed in the first part of the book have a strong communication component. Healthfulness, sustainability, authenticity and convenience all need to be communicated to the consumer if they are to give anybody a competitive advantage. Competences in food production therefore need to be supplemented by competences in communicating with consumers.

Klaus G. Grunert (ed.) **Consumer trends and new product opportunities in the food sector**
DOI 10.3920/978-90-8686-852-0_14, © Wageningen Academic Publishers 2017

This communication cannot be a one-way communication. While product positionings in terms of healthfulness, sustainability, authenticity and convenience need to be communicated to consumers, at the same time new product development in the food sector needs to be firmly rooted in communication from and about consumers, which is spread among value chain actors and can form the basis for the innovation process. Fortunately, the modern information society provides new channels of communication that can go both ways. E-commerce, mobile marketing, food-related internet fora, crowdsourcing of new product ideas all open up for new communication channels that can improve the flow of information from and to consumers.

In other words, the food sector is ready for 'open innovation'. The term open innovation has been expressed first by Chesbrough in his book 'Open Innovation: The New Imperative for Creating and Profiting from Technology' in 2006. Open innovation describes the change of creating innovations from a closed process, that goes on within the company and often has elements of confidentiality and secrecy, to an open process, where ideas and solutions freely travel across organizational borders (Chesbrough, 2006). We have mentioned the need for the farming sector and the processing sector to work together to bring about innovative food products. Likewise, there is some great unreleased innovation potential in the cooperation between food producers and retailers, a relationship that has been characterized more by conflict than by cooperation (Esbjerg *et al.*, 2016). And the best way to develop new ideas that match consumer requirements is the inclusion of consumers. The internet with its location- and time-independent social media networks is the engine of open innovation. So called crowdsourcing projects are an effective alternative to collect new input. Within announced ideas competitions knowledge, suggestions, experiences and desires are collected and evaluated. The innovator is using the collective intelligence of the mass.

Open innovation may also concern the open handling of data. Globally networked systems provide a lot of data that can be evaluated specifically to derive action strategies. The benefits of open data are expected to mount as more institutions adopt open access and data policies, and as more tools are developed and used. The responsibilities of those who own and use data are to not only develop new data tools and practices, but also to connect with other actors that can translate data into practical solutions and business models for improving agriculture and nutrition (Open Data Institute, 2015).

Some big players in the agrifood sector already manage large open innovation networks. In the field of processed foods, the larger companies and brands in particular are very active when it comes to improve or develop new products. On internet platforms such as TRND. com people register and can participate in product testing for free ('crowdtesting'). On the one hand, companies use the platforms for promotional purposes, to make new products more quickly known; on the other hand here products are tested before their commercial launch. In detailed questionnaires, the testers have to give a very accurate feedback. The data are collected centrally and can be used by the company for an accurate assessment. All everyday products are tested here. Especially high is the proportion of goods from the food sector.

Ideas can still be innovative, but if they are not funded, they cannot be realized. But this is less a problem for large companies, than for SMEs and private persons. In recent years thousands of ideas came true by financing them via crowdfunding. In crowdfunding, many internet users go together and provide equity capital to support projects, product developments or business models. So crowdfunding is a further step in open innovation.

References

Chesbrough, H.W., 2006. Open innovation. The new imperative for creating and profiting from technology. Harvard Business School Press, Boston, MA, USA, 272 pp.

Esbjerg, L., Burt, S., Pearse, H. and Glanz-Chanos, V., 2016. Retailers and technology-driven innovation in the food sector: caretakers of consumer interests or barriers to innovation? British Food Journal 118: 1370-1383.

Grunert, K.G., Fruensgaard Jeppesen, L., Risom Jespersen, K., Sonne, A.M., Hansen, K., Trondsen, T. and Young, J.A., 2005. Market orientation of value chains: a conceptual framework based on four case studies from the food industry. European Journal of Marketing 39: 428-455.

Open Data Institute, 2015. How can we improve agriculture, food and nutrition with open data? Available at: http://tinyurl.com/yact9eky.

Printed in the United States
by Baker & Taylor Publisher Services